Image by Design

Clive Chajet
Chairman and Chief Executive Officer
Lippincott & Margulies Inc

and
Tom Shachtman

Image *by* Design

From Corporate Vision

to Business Reality

ADDISON-WESLEY PUBLISHING COMPANY, INC.

Reading, Massachusetts	Wokingham, England	Sydney
Menlo Park, California	Amsterdam	Singapore
New York	Bonn	Tokyo
Don Mills, Ontario	Madrid	San Juan
Paris	Seoul	Milan
Mexico City	Taipei	

Library of Congress Cataloging-in-Publication Data
Chajet, Clive.
   Image by design: from corporate vision to business reality / by Clive Chajet and Tom Shachtman.
      p.   cm.
   Includes index.
   ISBN 0-201-55042-3
   1. Corporate image. 2. Marketing. 3. Industrial design coordination. 4. Packaging—Design. I. Shachtman, Tom, 1942– . II. Title.
HD59.2.C52   1991
659.2—dc20   90-46523
                                                                        CIP

Jacket design by One + One Studio
Text design by Lippincott & Margulies Inc and Melinda Grosser for *silk*
Set in 10.5-point Sabon by NK Graphics, Keene, NH

1 2 3 4 5 6 7 8 9-MW-9594939291
First printing, February 1991

To the heart and soul of my identity: Bonnie, Lisa, Lori.

C.C.

CONTENTS

ACKNOWLEDGMENTS

Adequately acknowledging all the people who helped me write this book would take another whole book, but I would like to especially thank some who were directly involved in the entire effort.

First and foremost, all my clients, who, unwittingly, have taught me all I know. Next, my many, many colleagues at Lippincott & Margulies including the late Walter P. Margulies, Gordon Lippincott, Ray Poelvoorde, Jack Weller, George Hafford, Irwin Susskind, Steve Gilliatt, and Jill Gabbe for her supportiveness and tireless energies and Richelle Huff for helping to make the book look so good. Also, very special thanks to Gershon Kekst for so generously sharing his insights with me, Pam Bernstein for her vision and encouragement, and Jane Isay for her editorial acumen and unflagging enthusiasm. Last, but not least, Wanda and Richard for feeding Tom and me so deliciously and Sandy Dein for her ever ready fingers.

Thank you one and all—identified and anonymous. I couldn't have done it without you.

Going to Work in Pink Pajamas

Every consultant has a nightmare. Most don't come true. Mine did as my colleagues from Lippincott & Margulies and I were in the Chicago boardroom of the National Can Company, making a recommendation for a name change that we suspected wouldn't go down well. Our conclusion was rooted in good research and logic, but I also knew why multimillionaires Nelson Peltz and Peter May, the major shareholders and leaders of the parent company of National Can, didn't want to convey this message themselves. Anticipating what was to come, I could hardly blame them for taking cover.

Peltz and May, chairman and president, respectively, of Triangle Industries, had hired our firm because when they acquired the National Can Company, the purchase had come with a catch to it. To land the prize, Peltz and May had reluctantly agreed that after the purchase, they would change the Triangle parent name and make it somehow reflect the parent's newest and most important asset, National Can. In the middle of our search for that new parent name, Peltz and May called a halt to it in order to share with us some inside information—that they were on the verge of buying American Can. They would then merge American Can with National Can and create the largest packaging company in the world. The deal was soon consummated and Triangle had a new identity problem: how to create an image that would hold together these former competitors under a single roof.

As we researched the matter, we learned that while National Can was better managed and more profitable than American Can, it had less of a positive image in its trade. American Can was seen as a high-class pioneer in research and development, responsive to customer needs, while National Can was solid but pedestrian. It seemed that the two images were mutually exclusive. Should we create a new name or use elements of both? Or neither? Should we choose one name over the other? Which one? After some soul-searching, we concluded that the better image ought to be as visible as possible and that the new unit should be called "American Can." The problem with this conclusion was that part of the corporate culture of National Can was hatred for the rival; each morning the executives had been told to go and beat out those nasty salespeople from American Can. But looked at strictly from an image standpoint, there was no contest: one should always go with the better image, even if it means problems. So we made our recommendation to Messrs. Peltz and May in the insulated comfort of their New York executive suite. They listened to our logic and conclusions, and then gave us the task of going out to Chicago and suggesting to the National Can honchos, who would run the combined unit in the future, that henceforth their unit ought to be known by the moniker of their former archrival.

It is sometimes our unenviable mission to be asked to convey bad news, and to protect CEOs from having to say things that they know will be unpalatable. Off to Chicago we went, in the dead of winter and between snowstorms, to make our presentation to the National Can executives.

In that drafty boardroom, I traced the logic of our research and came to the conclusion—that henceforth, National Can and its former competitor both ought to be called American Can. It was an electric moment. Hardly a beat went by before a senior manager said, and I quote, "That's the single worst recommendation I've ever heard."

I felt lower than a recycled garbage element, and thought that our delegation from Lippincott & Margulies might soon be tossed out into the snow.

We did, of course, have a fallback name: Trian Packaging. This, we told the National Can people (as well as Peltz and May), contained a combination of the *Tri* from Triangle, the *A* from American Can, and the *N* from National Can. That got a few votes, but didn't take the prize. The problem was that *our* image had been so tarnished by having made the original recommendation, neither side was open to any more original suggestions. The only name they could both agree on was the American National Can Company. A name at once logical and diplomatic, it's also quite a mouthful. It probably won't last, either: long corporate names invite shortening, and the likelihood, over time, is that this new company will come to be known simply as "American."

The final irony: Some months later, in what has become a legendary deal, Peltz and May sold their company for over a billion dollars to Pechiney Corporation, the largest French manufacturer of aluminum. The transaction required that Peltz and May buy back certain nonpackaging divisions, which then had to be bundled under a new parent entity. Since the Triangle name was weak—hundreds of companies use the word *triangle* in their names—a new one had to be found, one that, all parties hoped, would not run into legal problems anywhere. Since both Peltz and May had a passionate and superstitious attachment to the name that had previously brought them so much luck, they selected as the name of their new nonpackaging group— wouldn't you know it?—the Trian Group.

*   *   *   *   *

As can be seen from that story, working as a consultant in corporate identity and image management is exciting, but dangerous to one's health and ego. What we do gets to the heart of businesses—that's the exciting part. Most businessmen and businesswomen, though, think of image and identity consultants as people who sit around and come up with interesting names, smashing logos, and gorgeous design systems for corporations. But our work involves more than that, for the name of a corporate entity is but the keystone of a corporation's whole identity program—its look and its visual style. A good identity program must reflect the corporation's underlying reality. It all comes down to who you are, and how you present yourself.

In this book, I'll tell some stories about the image-related work that I and Lippincott & Margulies have completed for the hundreds of corporations that have sought our assistance. Though the details of the work Lippincott & Margulies has done on behalf of its clients remains confidential, I can reveal some of the thinking—heartaches—inventiveness—pitfalls— successes and failures that characterized it, and that embody general principles that all corporations and managers ought to know. The opinions offered on these stories are mine alone, and not always shared by our clients.

I came by the business in a way that sometimes seems to me roundabout, but, in retrospect, also seems to have been rather direct.

Born in London, I was educated at Columbia University, graduated with a degree in economics, and went to work first as an account executive in an advertising agency. I hated the job. An account executive was little more than the master of ceremonies who introduced the copywriter and the art director to the client and then was only occasionally able to intercede in the creative process. However, one of my first big accounts was West Virginia Pulp & Paper, a large manufacturer of packaging materials, and I became intrigued by the relatively new notion that good packaging could have a significant impact on sales, just as advertising was supposed to have.

Motivated by both my dislike for my job and my distaste for martinis—those really were the days of the two-martini lunch—shortly thereafter I went to work for a printer of packaging materials, Milprint, Inc., then a division of Philip Morris.

In those days, the job of representatives of such packaging companies was to roam the supermarkets and drugstores in search of poorly designed and generally lousy packages. Fortunately for us, there were lots and lots of bad ones. It was then necessary to dream up a better packaging idea, and, in a manner that would not insult the marketer of the dumb package (who was usually inordinately proud of his design), persuade him or her that your idea would be more effective. If the customer liked the idea, he was expected to switch suppliers and buy the printed packages from you. A number of potential buyers liked my ideas, but didn't like me nearly as much. Sometimes I'd go into the supermarket and see the idea I'd proposed, prominently displayed on a package printed by someone else—and would emerge feeling like I'd been raped. Ah, well; those were the bad old days. In any event, the only way my company and its salespeople made any money was by selling the printed packaging materials. We were, in effect, in the manufacturing business, yet saddled with the now outmoded necessity of having to give away design ideas and charging only for the packaging materials we manufactured and printed.

Although I was not a trained designer per se, I did have some creative design ideas and discovered that I was also a salesman; but I was selling mainly manufactured goods, rather than ideas. In the 1960s, most new packaging ideas were developed by materials reps or by the customers' internal design staffs. There were, however, some notable exceptions. Some trained designers who felt packaging design had a value independent of the cost of the packaging materials had made a business out of their design skills and talents, and were successful despite having to compete with firms that were giving away what the independent designers were charging for. Among the pioneers were Raymond Loewy, Walter Landor, and two particularly skillful and charming fellows named Gordon Lippincott and Walter Margulies. These men set the standards for the practice of packaging design, and spawned an industry. The firm of the latter two men, named Lippincott & Margulies, was rising to the top of that industry, and shortly became the leading packaging design company in the nation, and possibly in the entire world. The men's disciplined approach to the subject, their excellence in design, their relentless drive, and their talent for self-promotion made them stand out; they were the first with the most.

Since I enjoyed the creative-idea aspect of what I was doing better than the manufacturing side—I've always hated details—I decided to try to enter the packaging design business. I did not even consider applying to Lippincott & Margulies for a job, believing that there was nothing in my background that would interest a firm of that stature, since I had only limited experience in design.

So my next job was with Gould Associates, a small, Los Angeles–based packaging design firm striving to become national, for which I opened and operated a New York office. Jerome Gould, its founder, had a fine design portfolio that included the classic Michelob bottle. I stayed with Gould for a couple of years, during which I learned what to do as well as what not to do in the field, and ran some fairly large package design programs, the most substantial of which was for Pepsi Cola. Jim Summerall was president of Pepsi Cola in those days, and in a meeting he opined that if he were given the choice between the greatest package design in the hands of the most brilliant marketing executive, or a very long, very hot summer, he'd take the latter every time. This candid observation taught me that whatever I or anyone else was going to contribute to my client's prosperity, it was only going to be a part of the whole, and that more often than not forces completely out of my control were going to play a far more decisive role in the company's future.

Thus experienced in the ways of business, I yearned to

satisfy my entrepreneurial spirit by starting my own company. That opportunity arose when I met advertising legend George Lois, and persuaded him to join me in starting a design company. The Lois/Chajet Design Group got off to a good start, but eighteen months after we formed it, Lois's separate advertising agency had some financial problems that required his complete attention, and he was forced to withdraw. As the Chajet Design Group, the company continued to grow and became a market leader. Some of the well-known packages we designed included Ocean Spray cranberry juices, Sprite, Fresca, Maybelline, and CoTylenol. In the mid-1970s we bought a firehouse on Fortieth Street in Manhattan, and our headquarters came to symbolize the qualities that we thought defined our own company image: distinctive, stylish, memorable, and financially astute. Together with our clients, we developed marketing strategies; then our design staff would translate a particular strategy into the visual expressions that comprised the package.

This work brought us to the verge of the loftier field of corporate identity. However, we had some trouble convincing corporate executives that a packaging design firm could leverage its knowledge of brand issues to solve broader corporate communications problems.

Thus, in 1982, when I heard through a friend in London that Lippincott & Margulies might be for sale, I jumped. In the fifteen or so years since I had first become familiar with its operations, Lippincott & Margulies had metamorphosed from a packaging design firm to become the boardroom name in a new field. "Corporate identity problems?" a director of a company would say. "Get me Lippincott & Margulies."

Lippincott & Margulies had invented the field of corporate identity in 1964 and had dominated it ever since. In that year, while designing the Johnson's Wax pavilion at the New York World's Fair as well as some packaging for the company, Gordon Lippincott and Walter Margulies put some thought into a third problem facing their client. Johnson's Wax had just bought another company that sold a product entirely different from household waxes—health foods—and it occurred to Gordon and Walter that the wax image might damage the product line of the newly acquired company. Johnson's agreed that a potential problem did exist, and hired Lippincott & Margulies to look into it. Gordon and Walter called the process they undertook to solve the problem "corporate identity." Thus was the term and the discipline invented. It was the first time anyone had defined the relationship between image and identity. Gordon and Walter had come to the conclusion that identity was (in Walter's words) "that component of a corporation's image that can be wholly controlled by the company in its entirety."

Over the next two decades, the firm's services were sought out by many of America's giant companies, such as Coca-Cola, Chevron, Xerox, RCA, and American Express; as Lippincott & Margulies developed processes to fashion these corporate identities, it was also a pioneer in characterizing and influencing how a corporation's communications affect its interactions with the world.

In my own career, then, I had neatly recapitulated the progress of the corporate identity field itself, having moved from advertising to package design to the design of comprehensive identity systems. The next logical step was to become associated with the premier firm in the field, Lippincott & Margulies. When I heard that the company might be for sale, I called a very old and close friend on whose financial support and business advice I had relied from the beginning of my career, Leonard Stern, a perennial name on the *Forbes 400* list since its inception. He analyzed the balance sheet, assessed the reputation of Lippincott & Margulies in the industry, and said, with typical unwavering style, "Buy it—and I'll back you." I knew that Gordon Lippincott had long since retired, and that Walter Margulies was getting on in years. As it happened, I was then president of an industry group known as the Package Designers Council, an organization that, coincidentally, Margulies himself had founded, and so I had some reason to believe that Margulies would know my name when I called. He did know it, but at first denied the company was for sale until I mentioned the name of his lawyer (which my London friend, prominent packaging designer Michael Peters, had told me); then I was able to invite him to lunch at the Pinnacle Club.

Walter Margulies was a very polished man, elegant and charming, a European by birth, but one who always kept his background shrouded in genteel mystery. It was immediately obvious that his suave appearance masked the sharp, shrewd mind of a man who clearly understood what makes business tick, and the importance of profit to every business. I could see why the chairmen of so many leading companies had sought his advice, and, often as not, followed it. At lunch he was perfectly polite and listened to me patiently, but was noncommittal. After that, I didn't hear from him for a time that seemed so long that I prematurely wrote off the possibility of a deal.

Four months after our lunch, I received a call from Margulies' office asking me to come and see him again—though not in his office. During that four-month period, I later learned, Margulies had me checked out so thoroughly that I believe in certain respects he knew more about me than my wife did. Three meetings later, the deal was consummated, and I became sole owner and head of the company, with the understanding that Margulies would stay on for an unspecified period. Walter had ostensibly kept me away from the office during the negotiations so that his staff would not be upset by the sale of the company. Actually, as I later learned, it was because he was concealing a secret.

I discovered the secret at my first staff meeting: the company was a shell of its former self, with a very small staff, and only a few accounts, most of them inactive or in trouble. In buying Lippincott & Margulies at that particular point in time, I had bought only an image. My astonishment and panic quickly passed, because, in truth, I didn't care that Lippincott & Margulies was a shell; I would have come to work in pink pajamas, had Margulies required that, to be able to possess the name and the awareness level of the firm, and to have access to Walter himself, the founder of the field, who was going to stay on and tutor me. In this case, the image was not a shadow at all, it was everything.

And learn I did. Walter's genius for image making was manifest in many ways, small and large. For instance, he suggested that for ninety days the receptionist answer the phone "Lippincott & Margulies, Chajet Design Group," knowing full well that I would soon realize that the "Chajet Design Group" ought to be permanently dropped, since it in no way enhanced the prestige or broadened the appeal of Lippincott & Margulies. Similarly, he advised that I would probably decide to ease out of the packaging design business in six months, and he was right on that score, as well, although packaging design continued on as a corollary to our other work in image shaping. Our last single design assignment was the current Kellogg's Corn Flakes box.

In large matters, too, image—which is to say, reputation—proved beyond calculation. Six weeks after I arrived, an old friend of Walter's called to say that as a consequence of the government-mandated breakup of AT&T, smaller regional telephone companies were going to be formed, and all of them would need new corporate identities. Would Lippincott & Margulies like to handle the New York– and New England–based one?

Our contract with what became NYNEX was the turning point in the resurgence of our company, because we did good work and the program was highly visible. To this day, we continue to help manage NYNEX's image.

By the time Walter's cancer overwhelmed him, a year and a half after I'd come on board, I'd acquired considerable experience in the corporate identity field. Lippincott & Margulies was well on its way to employing as many people as in its earlier heyday. In 1986, I sold the company to Marsh & MacLennan, a NYSE-listed professional services firm generating $2.5 billion in annual revenues. This move not only benefited me financially, but provided Lippincott & Margulies with access to large financial resources and a global reach, within an atmosphere where Lippincott & Margulies has independence of operations, and we are encouraged to continue our pioneering consultant work. I remain as chairman and CEO, and we have a client list that reads like a ticker for the New York Stock Exchange—in short, a reality to match our image.

PART I

Business Problems—Image Problems

CHAPTER I

Business and Its Images

We live today in a world full of images, particularly business-related images. They are ubiquitous, from the still lifes of Andy Warhol's Campbell's Soup cans, to the NYNEX vans working on the corner, to the kids on skateboards whose T-shirts sport Coca-Cola logos. While we know and appreciate the power of such images as the Statue of Liberty and the American flag, we don't acknowledge in the same way the similar power of proprietary symbols such as the arches of McDonald's, though that icon of America is currently attracting a great deal of attention in Moscow. All around us are symbols, logos, and names created to differentiate one product from another in the marketplace; those identities, originally adopted for commercial use, have gone far beyond mercantile purposes—they have entered the lexicon and embedded themselves deep into our consciousness. As a consequence of successful image making, we travel not by train but by Amtrak; we go not to the grocery store but to the A&P; we don't leave home without American Express.

This book is about the role of image and identity in business, as well as the significance and impact these elements have on business. It explains how to understand these relationships, how to craft and evaluate images and identities, and how to manage them for profit and corporate stability.

As chairman and CEO of Lippincott & Margulies, the nation's oldest firm devoted principally to identity and image management, it is my exciting, rewarding, and often difficult task to help some of the world's largest corporations to understand, shape, manufacture, project, nourish, and maintain their own images.

Images All about Us

As critically important to corporate health as these processes are, they are not well understood by the general public, or even by a great many people associated with the large corporations, because few people really comprehend what is meant by image, especially in a business context. Here are a few recent newspaper and magazine headlines, all having to do with the word *image* as it is understood by the majority of the public:

"Jeffries Tries to Change Image"(Jeffries is an infielder for the Mets)

"Images of Women, Dignified or Not, but Always Nude" (photographic exhibit)

"Polishing Image, Principal Bans Gold Teeth" (such teeth do not convey the "serious image students need for the business world")

"Phnom Penh, Eye on West, Tries to Shed Image as Hanoi Puppet"

"Long-term Thinking and Paternalistic Ways Carry Michelin to Top ... But Uniroyal Deal ... Threatens Image as an Elite Producer"

"Computerized Stock Trading: Stormy Image"

Let me make clear the distinction between image and identity, since many people confuse these terms and use them interchangeably. A corporation's *image* is what is perceived by its various audiences—how it appears to outsiders such as the financial community or to potential consumers of its products or services; a corporation's *identity* is what it chooses to use to shape those perceptions.

Though today we live surrounded by images, there are still people—especially in business—who aver that they are unaffected by them. This is simply not true. Images affect us whether we want them to or not. Doesn't the Chrysler corporate image influence whether or not you will buy one of its cars? Doesn't the Goldman Sachs image add prestige to the shares it underwrites? Doesn't the visibility of Chemical Bank's identity enhance the idea that it is a safe place for your money? Wouldn't you rather work for IBM than for any other computer company? Buy the stock of AT&T based on its history and reputation, rather than on a painstaking analysis of its balance sheet? Like

Image by Design

it or not, our decisions in these instances are greatly affected by the images of the various corporations and their products.

Images influence thousands of our daily decisions, not only about business products, but also about politics and government; they shape our canons of taste. Image making has come to dominate American politics to such a degree that it is now not considered possible to have a viable candidacy without a carefully considered image.

"Newark's New Image Buoys Mayor on Eve of Election," says a headline in the *New York Times* (May 4, 1990). To erase the image of the city's 1967 riots, Mayor Sharpe James "has spent as much time over his four years as mayor trying to change that image as he has trying to solve the most serious problems." He installed around the city 775 nylon banners imprinted with flags and mottoes, then set out to alter the reality under the banners. He got the state's largest home builder to put up 1,100 apartments in the city, enticed Blue Cross & Blue Shield into moving 2,500 workers into a new downtown office tower—and did many other things that eventually resulted in an upgrading of city bonds, saving the city millions in interest charges. The ultimate gain, the CEO of New Jersey's Public Service Electric and Gas Company said recently, was a change "in the way Newark thinks about itself and the way others think about Newark." Image was at the heart of the transformation. Many a company under siege, or in need of regaining public confidence, could have used similar devices to boost employee morale.

Even foreign governments are concerned with image. Consider the problem faced by our client, the leading bank of the country of Turkey, which hired us out of frustration at having to combat the impression, conveyed in part by the movie *Midnight Express*, that Turkey was a backward and repressive society in which Westerners could not get a fair shake. Our task in Turkey was not made easier by the atmosphere in which we made our presentations in Ankara. All the members of the board of directors seemed to be heavy-set individuals who invariably wore menacing-looking dark glasses and listened to simultaneous translations of our comments with their eyes closed.

As to canons of taste: Donald Trump has become associated with a certain style of opulence, visible in the appointments of his Tower in Manhattan and his castles in Atlantic City, and perhaps in the styles of the shops that pay rent to him in Trump Tower. But his style has to do with extravagance: if marble connotes expense, he uses whole quarries of it in the new

Taj Mahal casino; but the effect, say critics, is more Disneyland than anything else. So "Trump style" is, at best, only reflective of a particular, rather limited, interpretation of luxury. But because Trump has been so successful at masterminding his image, his style greatly affects, and may well become, the majority's view of luxury, even though those who are actually very wealthy indicate that they regard Trump's style as, at best, rhinestone elegance.

Intrusions from Reality

Often, today, the outside world intrudes into corporate business in unexpected ways and creates problems, many of which have to do with corporate identities. Some are handled well, but the majority are not. Johnson & Johnson, faced with the Tylenol poisoning crisis that could have severely damaged the success of the company's many other products as well as Tylenol, was able to rely on its long-established and carefully maintained corporate image. But Procter & Gamble—a company with an equally sterling reputation, but one that had been less attentive to its corporate communications needs—underestimated the press reaction to a minor flap about its logo. When a rumor was floated that the P&G moon-and-stars logo was a symbol of the devil, it seems as if the solution P&G adopted to counter the accusation was worked up by the same committee that, when asked to design a horse, came up with a camel. P&G successfully sued the originator of the rumor, but also decided to remove the offending logo from its products, though not from its parent corporate communications. To me, the amazing aspect of the P&G caper was that a company that rigorously and regularly evaluated every other factor in its business, from the makeup of its products to its marketing strategies—ingeniously changing the formula for Tide and pioneering a revolution in baby care by convincing female customers to substitute disposable diapers for cloth ones—failed to reevaluate its corporate symbol in as systematic a manner in order to avoid problems *before* it came under attack. I assure you that no other asset in the P&G company is treated as casually as this one that is central to the company's image.

And what of the image problem that stuck to Exxon because of their handling of the Alaskan oil spill? Because Exxon viewed image as a factor relevant only in the selling of gasoline to consumers, and because it assigned a low significance value to its corporate image, in this crisis Exxon invited criticism. Exxon was perceived as carelessly flouting values held by a large number of Americans to be sacred. On the other side of the coin, when Mitsubishi of Japan (principally an industrial and financial concern) bought control of New York's famed Rockefeller Center, unleashing American sentiment that the Japanese are "buying up the country," statements issued by the Rockefeller Group about the integrity, commitment to excellence, and sterling business image of Mitsubishi quieted the furor after a few days. Those statements, added to Mitsubishi's obvious sensitivity to the value of an image—it paid a huge premium for the Rockefeller name and property—allowed Mitsubishi to successfully take up its position as a major new force in American real estate.

In short, in the current climate, corporate images are becoming increasingly and inextricably linked with a company's success and its future, both as a shield against troubles and as a potent weapon for growth. Corporate image, then, is a fundamental, mainstream management concern—one that has broad impact and implications for all other company activities.

Because we live in an age in which big business must bring in ever-larger volumes of sales through its products, firms tend to market only those products with the broadest appeal, products that are often very similar to one another: fundamental differences are slight between competing beers, stockbrokers, fast-food outlets, and travel services. Since an enormous amount of information assaults consumers every day, we are forced to find differentiators that will help us make decisions about our lives, decisions that involve far more than figuring out which car to buy. We need to decide where and how to live, what to do with our time, what hospital to trust for important surgery, how to insure our future. Yet the very elements that allow us to differentiate among our choices—corporate images, and image making itself—have been largely misunderstood. Images are thought to be mere puffery, tricks played on a gullible public. That's not so. One of our axioms—I can't count how many times I've had to say it to CEOs—is that *corporate images must reflect reality, not distort reality, if they are to have the most positive impact on the corporation's audiences.* Smoke and mirrors will not do; even if they are tried and seem successful for a short time, they eventually will be seen through, and will backfire on the company that has attempted to "get by" with them.

The need to understand the relationship between reality and image was evident in the *Wall Street Journal* interview (January 31, 1990) Peter Cohen gave on the eve of his dismissal as chairman and CEO of Shearson Lehman Hutton. Cohen forthrightly acknowledged the problems faced by senior management of the firm, but avowed that the solutions could be reached by "creating a new identity for the firm." (A loss of $950 million was announced a few months later.) His statement implied what many chairmen still believe, that a new identity can cover systemic faults. Nothing could be further from the truth. Conjuring a new identity to respond to mismanagement is never, never, never a solution—and in today's sophisticated business world, those who think it is will, like Peter Cohen, quickly become ex-chairmen.

An image is the imagined reality of its underlying subject, the one that is taken for the real thing by those looking at it. Good image-making marries the reality with the image, and bad image-making deliberately distorts the image in order to mask, protect, or otherwise prevent the accurate understanding of the underlying reality.

An image can be positive, negative, mistaken, accurate, overblown, too broad, or too narrow—there are hundreds of ways of characterizing images. The point is that an image of the corporation, and of its individual products and services, exists in the mind of its several audiences whether or not the corporation desires it to do so. That comes with the territory. The need, therefore, is for the corporate image to be shaped by a corporation's own management—or management risks having its image shaped by outsiders and even its competitors.

I define the work of the image consultant, then, as assisting the corporation in making sure that its audiences perceive the company as the company wishes itself to be perceived.

Let me say categorically that a lot of corporate images in this country are not being managed well. To name just one changed factor, many businesspeople have not yet acknowledged what to me has become obvious: that increased media attention and governmental regulation has made it impossible to do business without public scrutiny. That is to say, some factors affect today's companies whether they want them to or not, and these factors are often beyond a company's control. As a consequence, methods of doing business have had to change—a fact that some people haven't yet accepted.

A second instance; most businesspeople don't yet understand that when they get involved in a court case, two trials go on, not one: the first may be in the court itself, but the second is always in the arena of public opinion, and if the company wins in the first but loses in the second, it may have won the battle but lost the war. An example: Martin Sorrell, as chairman of WPP, bought the J. Walter Thompson and Ogilvy & Mather advertising agencies. At the time of the purchase, the smaller agency Lord, Geller, Frederico was a division of JWT. Dick Lord, the principal in that firm, was generally admired in the industry for such accomplishments as his corporate advertisements for IBM featuring Charlie Chaplin. Dick Lord didn't want to work for Sorrell, and initiated discussions to buy his small agency from WPP; eventually Sorrell decided he didn't want to sell, so Lord, Frederico, and twenty-five top employees walked out and started a new agency, Lord, Einstein, financed by Young & Rubicam. Sorrell sued Lord and other principals for a "conspiracy" to raid the old agency for its talent and its clients. There were two trials, one in court and one in the press. In the press, Sorrell was painted as greedy, unpleasant, and tough. In the court case, he won a judgment—but it didn't really matter, because 1) he lost clients, as IBM took its corporate advertising business away from WPP; 2) he lost a good agency, as the old Lord, Geller, Frederico firm is a shadow of its former self; 3) he lost credibility as a topflight, professional manager, because he hadn't managed his valuable image asset very well; and 4) he was perceived as a man for whom one should not work.

In the modern era, and for tomorrow, companies must pay increasing attention to what have previously been regarded as "secondary assets"—such concrete things as real estate, and such "intangibles" as employee spirit, goodwill, and other elements of corporate image. Actually, an image is something of great value to a company, if for no other reason than that it is totally proprietary to the manufacturer. It cannot be stolen, or outmoded, or lost to the Japanese. It can be protected by law, and so can the expression of it known as a patent or a trademark. Unless there is some fundamental change in the ways a company does business, the basic tools of image and identity—the logo and other communications items—generally last about twenty years. All too often, they are not reviewed until the end of that cycle, which is almost always too late for easy repair. Other aspects of a company's image, such as its advertising and its public relations, are reviewed annually, sometimes even quarterly. Corporate identity is much more permanent than transitory advertising and PR campaigns, but, all too often, executives are unable to separate out identity and deal with it in the ways that most affect corporate image.

All of image making needs to be better understood, both by the corporations who are the source of the images, and by us who are, more than we usually realize, consumers or target audiences of corporate images. This book aims to provide insight into the elements that make up an image; the ways in which images are perceived by various groups of consumers (from financial analysts to couch potatoes); how images are assessed, altered, or aligned; how they are disseminated and burnished; and how they relate to all the other aspects of the modern corporation.

Images are sometimes dismissed as impermanent, and more particularly as intangible and valueless; in this book I try to demonstrate—conclusively, I hope—that the impact of an image is so palpable that it can be felt, and so potent that it is often a company's most valuable asset.

By Choice, Not Default

Given that images exist whether we want them to or not, the problem in business is how to make conscious decisions about image, rather than letting such decisions be made by default.

An image is the sum of a corporation's conversations with society.

Corporations think they know precisely what they are saying to their various audiences, but often they don't. For example, a corporation has a plan and strategy for growth in the next ten years, and thinks that the outside world understands its direction and intent; but after making these plans public, management finds that the company's own P/E ratio (as reflected in the price of the stock) is lower than that of its competitor—and learns that that outside world doesn't see the company the way it wishes to be seen.

When we encounter such a problem, we ask the company: How do you see yourself? And: How do you wish to be perceived? Then we go to its audiences and ask: How do you perceive this particular company? Then we try to analyze the disparity among the answers, and in doing so we usually analyze the company's communications—its conversations with its various audiences—to determine what is helping and what is hindering it; and, of course, to try to figure out how to close the gap between where the company is and where the company wants to be.

This process of analysis goes to the heart and soul of a company. But most companies don't know that, and most companies don't want their souls searched—at least, not right away. In fact, most initially approach us with a request for us to design a new logo. In psychiatrist's terms, we might call that a "presenting symptom," because, quite often, when we look into a company's wants and needs, we find that the desire to change its logo is symptomatic of problems that are central to the company's business, problems that must at least be looked at, if not completely addressed, before we get on with the task of designing that new logo.

To nourish their images, all companies must and can do certain things:

1) As an institution, a company must have a clear foundation on which to build, a cultural foundation that takes into account the company's own history and strengths. What "worked" for the company while it was being built? Is it still working? When American Express came to Lippincott & Margulies, it didn't know whether its old name would suffice for the future it envisioned as a financial services giant. We convinced the company that its name was a strength that could be used as a foundation for greater enterprises.

2) Linked to the first item is the need to have a clear mission, preferably one that has been written out. What business is the company in? What are the priorities, the raison d'être? Where is the company heading?

If, in the 1920s and 1930s, America's railroads had recognized that they were in the transportation business rather than in the railroad business, they would now own all the airlines in this country, since they had ample opportunity and capital to buy up the airlines when they were small-fry. But they didn't, and were eclipsed. They hadn't put clear thought into the future, or, if they had, they certainly hadn't articulated a strategy based on insight or understanding of the direction in which mass transportation was headed.

In contrast, consider American Express. In the early 1970s, the company came to Lippincott & Margulies and laid a lot of its cards (no pun intended) on the table. The company had built up a travel service, a credit card business, and a service as a location to obtain mail in foreign countries, and it also owned Firemen's Fund Insurance Companies. Trying to analyze its own assets and looking toward the future, American Express thought it ought to head in the direction of financial services. But the executives were concerned that the company's name and, consequently, its image, were perhaps not appropriate for this goal.

If only American companies were as thoughtful about their future missions! But most are not, and so have somewhat of a difficult time accomplishing the third objective, which is to . . .

3) Adopt identity practices to align the audiences' perceptions of the company with what the company believes itself to be, and the goals it hopes to accomplish.

*   *   *   *   *

In the next chapter, I'll present and analyze a fictionalized worst-case scenario of a company that in its ignorance about the connection between business planning and image blithely ignored all three of these key image-nourishment objectives.

CHAPTER 2

TPI Doesn't (I Hope) Always Stand for Typical

In January 1989 Lippincott & Margulies put together a seminar for the Conference Board in New York City on bringing corporate identity into focus. Now most people, if they think of image and identity consultants at all, believe them and all who sail with them to be name changers and graphic designers, pure and simple. One of my objectives in chairing this rather unusual presentation was to demonstrate the broad range of problems in which Lippincott & Margulies gets involved *before* a name change can even be contemplated. To demonstrate, my associates and I conjured a case history for an extremely unlucky company we called TPI, Inc., and introduced a series of vexing problems facing it—which I, as chairman and CEO of the fictitious company, would have to face along with my board of directors, who were to be the panelists for the discussion. Any resemblance to any real companies, living or dead, is purely coincidental, as I'm sure you'll agree after becoming privy to the tale and the travails of TPI, Inc.

My board included corporate communications executives, public relations and identity consultants, and two well-known financial journalists. In alphabetical order, they were Kim M. Armstrong, Director of Consumer and International Long Distance Advertising, of AT&T; Karen Bachman, Vice-President, Public and Financial Relations, of Honeywell, Inc.; David R. Butler, President and Managing Director of Lister,

Butler, Inc.; Rance Crain, President/Editorial Director of Crain Communications, Inc.; Carlton L. Curtis, Vice-President, Corporate Communications, of The Coca-Cola Company; Roxanne J. Decyk, Senior Vice-President for Administration of Navistar International Transportation Corporation; Richard C. Hyde, Executive Vice-President and Managing Director/Corporate Relations of Hill & Knowlton, Inc.; Franklin T. Jepson, Corporate Vice-President, Communications and Investor Relations, of Bausch & Lomb; Myron Kandel, Financial Editor of Cable News Network, Inc.; Joseph M. Kayal, Director-Corporate Identity, NYNEX Corporation; Edward J. Meyer, General Manager, Marketing, for Sun Refining and Marketing Company (Sunoco); Al Ries, Founder and Chairman of Trout & Ries, Inc.; Leonard H. Roberts, President and Chief Executive Officer of Arby's, Inc.; Claude Singer, Vice-President, Director of Policy Communications, for Chemical Bank; and James A. Taylor, National Director of Marketing for Ernst & Whinney International. Every CEO should be as fortunate as I was to have such a sterling board.

To set up the problems, we presented an encapsulated background history of TPI, Inc.:

Texas Oil & Petroleum Products, Inc. was founded in 1907 by a poor cattleman who didn't stay poor very long once oil had been discovered on his land. Very quickly the firm became one of the world's largest integrated petroleum companies, involved in discovery, drilling, refining, and marketing operations around the world; its Starbrite oil products are retailed through service stations in twenty-three states and four European countries. In 1988, TPI reported $500 million in profits on total sales of $50 billion. It has ten thousand employees and is globally based and publicly traded, with management owning less than 5 percent of the stock and a shareholder base heavily weighted to institutional investors.

The company has been diversifying for a half-century. In 1924, during a slump in petroleum prices, it began farming vegetables on part of its vast land holdings; soon the Texas Foods Division had become one of the largest vertically integrated food processors, a leader in the manufacture of baby foods and frozen vegetables; food products accounted for 28 percent of total revenues in 1988. In 1976, the company decided to diversify again and purchased a small Detroit-based company called Cellular Fiber Optics Laboratories. At its Detroit facility, Cellular has a model hiring and training program for inner-city minorities, called Work Is Now (WIN), which includes a division-owned and operated day-care center that enables welfare mothers to hold full-time jobs. Cellular itself grew so large that it bought up dozens of mom-and-pop paging companies across the country and became the nucleus of the nation's first national telephone messaging and paging network, SateLink, Inc.

In 1978, reflecting the fact that it was no longer exclusively tied to the petroleum business, the company changed its name to TPI, Inc. In a last move in 1982, TPI purchased a company called Financial Futures, Inc., and turned it into a top asset-based lending institution that loaned funds secured by the pooled receivables of hundreds of midsize businesses and operated as a division called Texas Financial Services, Inc.

The board recently decided that the company needed a major overhaul of its corporate identity practices, including, perhaps, a change of name. As so often happens when a company gets to the point of considering a name change, TPI must address some difficult underlying problems. The board and I had every intention of getting to those, but before we could, a disaster struck and demanded our immediate attention.

Baby Treet Gone Bad

A month ago, a jar of Texas Foods' Baby Treet strained peaches was found to contain arsenic. One child in Boston was taken ill, but fortunately recovered. We recalled all batches of the peaches with the same production code. Just yesterday, however, we received a report that three babies died in Illinois after eating Baby Treet strained peaches. Clearly, I told the board, TPI, Inc. now faced a serious image and communications problem. What ought we to do about it?

Everyone agreed we would immediately have to recall the product, without first taking any time to find out what had gone wrong or to assess blame. Carlton Curtis of Coca-Cola suggested we could think about reintroducing it "only after consumer confidence had been restored." AT&T's Kim Armstrong said we ought to withdraw all advertising and sales promotion for the product and hold a press conference to tell the world what we were doing and why, and what we knew about the problem so far. Rance Crain insisted that TPI must be forthright and open with the media, for if they felt we were withholding information, they would attack us and never let go. Franklin Jepson of Bausch & Lomb pointed out that the CEO of TPI—not the head of the Texas Foods Division—must step up to the mark and personally reassure the public of the company's actions in response to the crisis. David Butler cautioned, however, that the CEO must distance himself from the technical problems that had to be solved by the foods division, though others weren't so certain it would be possible to insulate the corporate parent from the problem that had overtaken one of its major divisions. Everyone joined together in hoping that the company's corporate culture would be strong enough, and based enough on integrity, to help the company withstand the crisis; that's what had happened when Johnson & Johnson was faced with the matter of poisoned Tylenol.

Bond Slippage

Before we could complete our discussion of the baby food crisis, I had to delve into another problem: Standard & Poor was about to lower its Triple A rating on our bonds because of questionable investment practices of senior officers of Texas Financial Services; even worse, 60 percent of the bonds were held by the TPI employees' pension fund, and as word leaked out about the lowered rating, employees were becoming alarmed. My first question to the board was what ought to be done to calm employees.

Roxanne Decyk of Navistar insisted that the facts be laid before employees promptly to quell their nerves, but Al Ries warned that to send telegrams to employees telling them to disregard what they'd read in the newspaper would simply alarm them more. Arby's CEO, Leonard Roberts, said that no matter what, the correct information must come from management to the employees, not from outside sources, or the management's credibility would be shot. Edward Meyer of Sunoco said that since his drivers always seemed to be the ones who knew what was going on in the company before anyone else, TPI ought to fully explain the bond matter to drivers and other employees who had direct contact with suppliers and customers. This tactic, he said, would help spread TPI's point of view that the downgraded rating was a mistake and that the company was of underlying sound financial health.

I asked whether internal and external audiences of the company must be considered with equal emphasis in this crisis. Joseph Kayal of NYNEX believed adamantly that if our internal audience—our employees—weren't properly convinced of the wisdom of the company's actions, no attempt to shore up outside credibility could be sustained for very long. Claude Singer of Chemical Bank pinpointed the most important part of dealing with that internal audience: if the corporate culture trusted us in management—for instance because we'd dealt with the employees honestly in the past—they'd be likely to believe us now; and if not, there'd be hell to pay. A company needs a corporate culture to get it through a crisis like this.

Mike Kandel upped the ante by pointing out from his perspective as a financial journalist that the story of the lowered bond ratings could not be contained; journalists would smell blood in the water and subject TPI to "an ongoing barrage" of questioning. He warned me that I, as chairman, must be prepared for it. Karen Bachman seconded this horrifying notion, advising me to get all of the financial bad news out at once, rather than in dribs and drabs, and to try to attribute that bad news to Texas Financial Services, rather than letting the blame spread to the whole TPI organization. I wondered to Claude Singer whether I should publicly fire the head of the financial services division to take the heat off TPI. Singer agreed it had to be done, and Rance Crain suggested the division head be replaced by some well-known financial figure considered beyond reproach—someone who could bring instant credibility to a troubled institution.

Would that be enough? The lowered bond rating, even more than the baby food poisoning, would cause headaches for TPI top management for some time to come. Frank Jepson said the present troubles were an indication of the deep difficulties facing the institution. Two crises, each in a different division, showed that the company was getting out of control. Perhaps it was starting to unravel because it had diversified too fast.

I was just getting around to saying that all the panelists/advisers seemed to agree that a good corporate culture was a great asset, when the onstage phone rang.

Win or Lose?
In this phone call, I was supposedly speaking with the head of TPI's Cellular Division, who informed me that because of a downturn in business, he was going to lay off six hundred workers in Detroit. Obviously, he said, the WIN program had to be suspended since there were no longer any job openings; and since two hundred of the workers were mothers of the children attending the WIN day-care center, he was also planning to close the center. My division head informed me that he feared that the fired minority workers would file a class-action suit charging Cellular with de facto discrimination against welfare recipients.

Mike Kandel categorically labeled this a "major management blunder, a public relations disaster." He reminded me that TPI's last annual report had included three pages of pictures about the WIN program, and that it had been featured on his own network, CNN, and on CBS's "60 Minutes." Hyperbole notwithstanding, Mike pointed out that TPI was now allowing a manager at the local level to unilaterally shut down the WIN program, and that this could spell unalloyed disaster.

The board offered several suggestions for heading off the crisis, two of which were quite innovative. The first was that WIN could be refocused to become a retraining and outplacement organization designed to help Cellular's former workers negotiate a "soft landing." The second was an attempt to turn a negative into a positive by having Cellular's manager meet with other Detroit-area companies that were experiencing a problem with minority hiring and propose that WIN become a jointly sponsored endeavor. Whatever the solution, my advisers agreed unanimously that the corporate parent ought to take over this nationally known program, for it could be kept going at a relatively minor cost; but if it were precipitously dropped, the cost could be enormous.

The advisers were adamant that decisions of this magnitude—ones that could affect the parent company's image—ought not to be made by a subsidiary without prior consultation with the parent. There was a need, said some, for a centralized communications facility at the parent level, one that would supersede the autonomous operations of the subsidiaries. Others disagreed, saying that centralizing communications would be tantamount to a reorganization, a major project that shouldn't be entered into lightly.

Some of my advisers were so brash as to suggest that this third instance further demonstrated that TPI was out of control, and that top management and I hadn't been properly minding the store. Couldn't we keep our middle-level employees in line? Dick Hyde of Hill & Knowlton suggested it would be a good time for the board of directors to step back from specific problems and address the matter of where the whole TPI company was going.

Stepping out of my TPI role for a moment, I told our audience I was most pleased to see that the knotty problems facing the fictional company were all being assessed and addressed in terms that had to do with corporate communications.

Rumors of Offers

Next problem: an LBO firm wants to meet with me to discuss an unwelcome leveraged buy-out by some shareholders who say that TPI, broken up, is worth 40 percent more than its current stock price suggests. TPI must now consider selling some of its assets to stay afloat. I put to my advisers the question of whether we should sell the Starbrite gasoline stations and brands, and, if so, how we could decide the true value of the Starbrite brand. Carlton Curtis of Coca-Cola acknowledged that a brand's value was hard to assess, but confided that the accepted way of guessing at it was to figure out what sort of time and money would have to be invested to build a similar brand and share of the market from its inception; that is, what it would cost to fund a replacement brand. While Al Ries cautioned that the brand might actually be worth nothing at all if consumers had no loyalty to it, David Butler disagreed, pointing out that "the major brands today are substantially the same ones as in the 1920s," and that since ours was an old company and Starbrite a well-established brand, it could be assumed to be worth a great deal of money.

Kim Armstrong of AT&T pointed out that TPI should first determine "what business we're in" before evaluating the worth of the business's individual pieces such as Starbrite. If, for example, we chose to get out of the gas station business, Starbrite could well be assigned a lower value than if we decided to go back to our original strengths, sell other sorts of divisions, and concentrate on a core gasoline enterprise.

"If our objective is to remain independent," I asked, "what do we have to do vis-à-vis the financial community to reassure them that we could or should stay independent?" Frank Jepson said that because of all the blunders and catastrophes, TPI had actually lost the right to simply insist to the world that it was and should remain an independent company. Control of its fate was shifting into the hands of arbitrageurs, who would decide for themselves what it was worth. Board members agreed that we ought to immediately create a committee composed of company directors and some outsiders, which could assess the breakup value and decide what sort of reorganization might give the company a fighting chance to stay alive. I summed up the advisers' sentiments as follows: the corporation needed identity practices and communications to show that it had a reason for being and a value worth preserving.

Restructure and Rename

Because of the four crises, management decided that the company ought to be restructured and slimmed. For purposes of discussion, we had already hashed out the new organization. TPI would now have three major divisions: 1) "communications technology" such as Cellular and SateLink; 2) "energy exploration," including Texas Oil and Starbrite; and 3) Financial Services. We would divest the Texas Foods and TPI technology divisions. To go along with the reorganization, we also decided to overhaul our corporate identity practices. To communicate our new image, ought there to be a new corporate name?

Kim Armstrong at first thought the company's divisions could be named by categories, something on the order of TPI Financial Services, TPI Gas & Oil, TPI Communications. "But how much does the name TPI itself bring to the equation?" she wondered, and then partially answered her own question by pointing out that it hadn't helped the stock price—in effect it had not been a strong brand name.

It was interesting to see that even here, in this playacting seminar, many panelists objected to changing the name of the corporation—just as real boards of directors very often do, and on similar, emotional grounds. Joseph Kayal of NYNEX thought there still might be a lot of residual equity in the TPI name, and Leonard Roberts of Arby's said that since TPI was the historic unifying principle behind all of the current divisions, we ought to think seriously about continuing to use it. Roxanne Decyk wanted to delay the entire process, suggesting—quite rightly— that the company and its management were under a great deal of stress and strain at the moment. In a year or so, when we had addressed the serious problems brought to light in the four crises of the day, we might be able to tackle the issue of a new name; as of now, it was too threatening.

I needled Rance Crain, a frequent critic of corporate name changes, and of mine in particular—he once called me "Public Enemy No. 1," a step up from his former designation of me as "my old nemesis"—to come up with some good ones. To sum up what Rance said and thought just then is unnecessary, for he later put it in his column in the March 13, 1989, issue of *Crain's New York Business*, part of which I'm pleased to include here. After giving the background on the seminar and on TPI's escalating crises, Rance complimented me on my handling of the various advisers, and then headed into dangerous territory.

> It's not easy for me to be so gracious when it comes to the man who has been responsible for many of the more notorious corporate name-change atrocities of our time—including NYNEX, Hartmarx and Allegis—but I must say he has taken my editorial fulminations on the subjects in the constructive spirit in which

they were intended. It's my deeply held conviction that too many corporate name changes have contributed to miscommunication by dismantling the very essence of what a company stands for and erecting in its place a computer-derived name that evokes cold and sinister connotations.

The strangling of language and names with the help of computers has been going on for some time, of course, but it's been fairly gradual and unobtrusive until the last few years. But ever since the dealmakers have taken over this and sold that, they've looked for something alien and heartless to call their conglomeration, constructed to convey a single-minded, unwavering march toward a forever upward bottom line. So they concentrate on concocting a name conjuring up dollar signs to the financial public and money managers and never mind what damage they may do to their other constituents, such as consumers and employees.

But I was forced to make an exception with the company that seemed to have fallen on irreparable hard times, TPI, Inc. My thinking in this case was that a name change might be the only way for the poor company to start afresh. So when my turn came to recommend a course of action, I opted for a name change to Finopto, S.A.—"Fin" for financial services; "op" for cellular optics, and "to" for Texas Oil and Petroleum Products. My thinking was that the name has a foreign ring to it, which is good so nobody would think that an American company was responsible for so many things going wrong. The S.A. at the end, which many foreign companies actually use, actually stands for "We'll sell anything."

This town's big enough for both of us, Clive Chajet. I'm beginning to get the hang of it.

—Rance Crain

CHAPTER 3

How Is Your Business Perceived?

Business today wants its audiences to think well of it, and this is a matter of images. *An image is the imagined reality of a subject.* That is, it is not necessarily the reality, or even a precise mirror of it, but, rather, what people believe to be real or want to believe is real. This is especially important in business. Let me list and briefly analyze the elements that go into our perceptions of businesses through their images. The purpose of the list is to catalog the elements; the danger is that some people may treat the list as a Chinese menu from which to choose only certain items that one deems especially succulent—or relevant. My point is that all are relevant, and all are interconnected. There is no precise formula for optimum balance among the elements. We only know that all to some degree contribute to the way a particular business is perceived by its audiences.

How the Product or Service Performs

A company's image, or that of one of its brands, depends for the most part on how good the actual product or service is. Since many people believe image consists mostly of smoke and mirrors, my insistence that image depends on performance may seem surprising. But reality is at the core of an image. Delivering on the promises made by a company's advertisements—whether for good-tasting burgers or for solid investment advice—promotes public acceptance of the promises. *Caution: Since a good image is almost always the reflection of solid performance by a company's products and services, no other single factor can be as important in producing a positive image.*

The Price

The image of a company or a product bears a direct relationship to the price paid for the company's manufactured goods or services. The higher the price of a product or service, the better the product or service is perceived to be. It is well documented that people pay more for quality. What is less well known, but no less true, is that when they pay more, they also *assume* that what they get for their money is a better product or service than if they'd paid less for it. That's not always so, however, because there are two different ways to price a product or service; one is to calculate costs and add a normal profit margin to arrive at a price; the second is to begin with the notion that a certain price level is desirable for shaping a particular image objective, and to build upward from costs (not always in a precise multiple of those costs) until that level is reached.

Thus the fees McKinsey & Company charges for its consultancy studies and recommendations are not simply a function of the costs, but are more a reflection of the seriousness with which clients view those services. McKinsey charges a good deal, and clients generally feel that anything costing that much must represent valuable and useful advice. In other words, McKinsey is able to use its price as an image-building device for its consultancy practice.

You might expect that people do not like to overpay and would object to unfairly high prices for goods or services, but this is not always so. Australian wheeler-dealer Alan Bond spent $59 million on a Van Gogh painting because he presumed that the purchase would propel his own image into the stratosphere.

Conversely, a compelling case can be made that lowering the prices of products or services that have traditionally been expensive can hurt those products or services. For example, when Avon owned Tiffany, it decided to market a line of gifts at Tiffany's that could be bought for very little money, for instance a $7 wineglass. Prior to that time, the only item in the store that cost less than $25 was a fancy toothpaste-tube roller. The new, low-priced line sold well, but severely damaged Tiffany's core business, which consists of ultra-expensive jewelry, silver, watches, etc. You certainly wouldn't want to buy a $100,000 necklace bearing the same name and packed in the same sort of classic blue box at the same store from which you could purchase a $25 pin, would you? Many customers did not.

After the firm was repurchased from Avon, it went back to the high-priced merchandise that had made it famous. And the management capitalized on the name and cachet in new and expensive products. "Nobody buys the Tiffany perfume for its scent," said a marketing researcher (*Forbes*, February 6, 1989). "They want that name and blue box."

It is possible to offer a relatively low prices and suffer neither loss of profit nor a damaged image. The heart and soul of Chevrolet's image strategy is its reasonable prices, and General Motors makes sure that the pricing policy for Chevrolet always upholds that image. Of course, if costs rise, Chevrolet has to respond by raising prices, knowing that in doing so it risks altering its image. The only consolation is the sure knowledge that if its own costs are going up, so must its competitors, who must then raise their prices as well; thus, the relationship of Chevrolet's price structure and image to that of the competitor will probably not drastically change.

A recent survey by a marketing consultancy called The Public Pulse reported in September 1989 that the United States is having trouble marketing high-quality, high-priced merchandise—to Americans. Research conducted in six hundred households with high annual incomes showed Japanese goods receiving the highest marks for "best value for dollar," "highest quality," "most innovative," and "most reliable," outranking those of the United States and Germany; goods made in France, Great Britain, and Italy finished in the back of the pack. The report concludes that, based on their success in moving products "upmarket"' into luxury niches, the Japanese have laid a tremendous foundation on which to build for the future. American products will have to fight to regain first place in the field of high-priced, high-quality goods in the coming years.

Caution: Do not expect that a high price will by itself bring a luxury image to a product or service. If customers perceive a high price as being just too much to pay, failure is almost guaranteed. A $50,000 price tag nearly stalled the Allanté Cadillac; car buyers simply weren't prepared to pay that amount for an American car. While Cadillac originally thought it would be an image advantage to be able to claim that Allanté was the most expensive American automobile, the high price turned out to be an insurmountable disadvantage.

The Name

Whether it's a name of a company or the name of a particular brand, this capstone of an identity is vitally important to how a company or product is perceived. Some names connote importance or dependability or luxury, others suggest low prices or quick availability or no-frills service. All names need to be appropriate for the company's purposes, and to convey those purposes with clarity. In the 1970s, a company called Family Bargain Stores owned a number of subsidiaries, including Cartier, Mark Cross, and Kenneth Lane—three "luxury" names that didn't fit the parent company's name or image. In fact, were Cartier, Mark Cross, or Kenneth Lane to be associated in the public's mind with cut-price bargains, the result would be devastating. Knowing this, the parent company adopted the name Kenton Corporation, which sounded appropriately neutral and suggested that the company was a pillar of the establishment; fortunately, it could also be rationalized since it was derived by combining the names of the two entrepreneurs who were the driving force of the organization.

Sometimes the owner's name is no longer adequate for new business purposes. The Mallory Battery Company had a good product, but the name Mallory just didn't help it. When the company consulted us, we reasoned that the name ought to signal a consumer benefit, giving people a reason to buy it. We created and recommended Duracell, a name that connoted a long-lasting product and could be used as a marketing tool. In fact, the name immediately distinguished the company from its competitors, and Duracell seized a larger share of the alkaline battery market.

In some cases a name must help sell a product or service that otherwise might not be salable. America's largest manufacturer of coffins, Hillenbrand Industries, had decided to offer a new service. Its research showed that while most people dislike the subject of death, the issue of funerals is even more distasteful. Heirs and surviving family members often experience great pain when they have to arrange the details of a funeral of a loved one. Hillenbrand believed there was a market for a service that would enable individuals, in advance of their death, to arrange every detail of their funeral, thereby relieving survivors of a difficult burden. Now, what should this service be called? As we looked into the matter, we concluded—to our initial surprise—that the most appealing feature of this service was not that the cost of the funeral would be paid for before death, but, rather, that survivors would be spared the pain of having to arrange and pay for the details themselves. A customer who purchased this service would be acting more out of compassion than a desire to contain costs. Once we understood this, we came up with the name Forethought. The service has been phenomenally successful, and while the way in which it is administered has been the major factor in its success, we like to think that our correct handling of the image issue, through a particularly appropriate name, has also been of vital importance.

Caution: Don't place too much of a burden on a name. Good names are hard to find, and most dictionary words have already been registered for use by some firm or other. Then, too, most generic descriptions are not distinctive enough to serve well as names, and many contrived words are difficult to read. While a great name has the potential for being the single most effective tool for image shaping, such names are hard to find.

Business Practices

How the company's business is run is a major determinant of how it is perceived. This is true not only in regard to the company's pricing policy, but also to the way in which it runs its operation. When Gucci came to the United States and opened the most expensive shop on Fifth Avenue, it closed that shop for lunch from one to two, conveying the message that Gucci was no ordinary retailer, and that it didn't want as customers secretaries who had only an hour (and not much money) to shop. Gucci was for the well-to-do woman who could arrange her own shopping schedule to fit the store's hours. Going one step beyond Gucci was the exclusive men's shop Bijan, whose two locations are on Rodeo Drive in Beverly Hills, and on Fifth Avenue in New York: it accepts customers only by appointment. No off-the-street traffic for Bijan.

On the other end of the scale, consider 7–11. Its business practices are also quite distinct, and no less successful—though the goods it sells are considerably less dear than Gucci bags or a Bijan suit. The message conveyed by the name as well as the business practices of 7–11 is that the stores are open for the convenience of customers. The 7–11's open earlier and close later than do conventional supermarkets; that is, they sell convenience, and ask customers to pay a premium price for it.

Consider the interesting case of how some products that purport to be American—but aren't—are sold in Japan. A chain of fifty-six Aunt Stella's cookie shops is very successful, and customers think that both the chain and its sweet-smelling products are American; but there is no such chain in the U.S., and

American consumers might be put off by the extra-sweet smells that the Japanese find so alluring. The business practice is effective in Japan, though, and that's what counts. According to a recent article in the *Wall Street Journal*, American images that sell well to the Japanese include those of the Colonial era, the 1950s, and power-hungry New York businessmen; one doughnut-store chain features pink neon flamingos, revolving merry-go-rounds, and music from the 1950s (complete with comments in English from a disc jockey), all of which feed on Japanese nostalgia for an era when, the chain president says, "America was our dream."

Caution: When selecting business practices, make sure they reflect sound strategies more than they do a concern for image. It's important to pick a strategy because it will work, rather than because it may look good. That having been said, however, image must be considered as a factor in most such decisions. Employees who are paid only minimum wage by a skinflint employer will service the company's customers only minimally. But the sword cuts both ways. When customers fly on what they know is a discount airline, they don't expect the same sort of service from what they assume is a low-paid flight attendant that they might insist upon from a better-paid attendant on a full-price airline. The equation between business practices and image is always there, and policies must be adjusted to take it into account.

The real danger arises when the image component of business practices is ignored. And the real profit may be reaped when that image component is strong enough to outweigh a glitch, a mistake, or even an accident in business practices: Source Perrier, the largest bottled-water purveyor in the United States, is a case in point.

When some bottles of Perrier were found to be tainted with benzene, Perrier immediately recalled all product from shelves in the United States, and then from shelves all over the rest of the world. Some analysts thought the company overreacted; others commented that if the product was off the shelves for longer than a few days, Perrier would lose market share and have a hard time regaining it. I didn't think so, and neither did Perrier officials, who understood that total recall was essential. "I have built up this company over the past forty years around an image of perfection," said the chairman of Perrier at a news conference in Paris; "I don't want the least doubt, however small, to tarnish our product's image of quality and purity . . . Our image and the respect and confidence of consumers have no price for me." I couldn't have said it better myself: here was a man who recognized that total recall was necessary to reinforce the image of purity on which Perrier's phenomenal sales had always been based. By removing the potentially tainted bottles, finding and fixing the problem, and refusing to ship new bottles until purity was assured, Perrier was reaffirming the basis of its business, a convenant of purity that bonded company and consumer.

I thought then that if Source Perrier resolutely clung to its historic strategy of marketing the product as a pure beverage and took care that no impurities were allowed in it, Perrier should in time rebound and realize a full restoration of the company's most valuable asset, its image. And, of course, the company would receive priceless free publicity at the time of the reintroduction. Although the sale of the certain company's brands in the interim between recall and reintroduction somewhat clouded the issue, the image of purity seems to have carried the day.

The "Look" of a Company

How a company appears—the visual expressions of its identity—determine in large measure how the company and its products are valued by its customers and audiences. The public sees logos and other visual elements of companies every day, and though surveys show that we think we don't pay much attention to them, we do, we do. Sometimes a logo takes on a life of its own, going beyond its function as a symbol appearing on a letterhead or at the bottom of an advertisement to the point where it becomes the heart and soul of a company.

Take Mack Truck, for instance. In the late 1980s, the company went into a deep decline. In an earlier time, Mack Truck had embodied American industrial power, symbolized through its ubiquitous bulldog logo, which appeared on the grilles of the massive trucks and on lapel pins that were given to visitors on their way to lunch in the bulldog executive dining room. As part of an effort to reverse the downturn, the company brought in a new management team, whose members understood the strength and symbolism of the logo; in discussing plans for the future, new CEO Ralph E. Reins invariably used the phrase, "The bulldog is not going the way of the dinosaur." Nothing more needed to be said, and the logo became a rallying point for the company.

When two IBM computer experts made a breakthrough in miniaturization, how did they choose to demonstrate it? By producing the world's smallest corporate logo, a symbol whose length was measured at 660 billionths of an inch. When an American was appointed head of the American operations of the Japanese securities firm, Nomura, he suffered from low status until precedent was broken and he was awarded a company lapel pin in a ceremony at the graveside of the company's founder. Clearly, the lapel pin was much more than costume jewelry.

Don't expect too much from a logo though. A logo can frequently be an effective identifier that distinguishes a company from its competitors, but it may not prove particularly useful in shaping the company's image. There's nothing wrong with the sort of logo that fulfills limited goals, but the difficulty arises when a company allows its logo to be so distinctive that its effectiveness becomes diluted. For me, the Chase Manhattan Bank logo, an abstract design, falls into this trap, as does Citibank's star. Citibank's situation is even worse because some of the bank's units have adopted the distinctive shape of the company's Manhattan building as their logo. Had the Citibank star been more meaningful, unique, and relevant, such unauthorized experimentation with other Citibank logos probably would not have occurred. Similarly, the Goodyear symbol, with its winged sneakers, sends a confusing message to those who might wonder if it stands for a company that sells sneakers rather than automobile tires.

Quite often, symbols are used incorrectly. Borrowed classic clichés bother me in this regard. Why would European coats of arms, shields, or family crests be used to sell Canadian whiskeys or American cigarettes? Does either category of products want or need a European aristocratic image? Seagram's 7, a uniquely Canadian product, was originally positioned as an alternative to Scotch whisky, and Marlboro cigarettes are as American as the cowboy who so effectively dominates their image—but both have borrowed crests. Why? Beats me.

Take another seemingly innocuous element, the corporate colors. A minor matter? Not at all. Would IBM conjure a solid image if its colors were pink and lavender instead of its corporate blue? IBM, nicknamed Big Blue, would suffer more than a color change if its nickname became Big Fuchsia.

Considering the prominence of green in its packaging and other materials, could one imagine Salem cigarettes being anything other than mentholated? Would Campbell's Soup convey the image of full flavor as successfully if its label were pale yellow and white instead of red and white? Miller High Life has always had difficulty convincing beer drinkers that its brand is full strength, because it comes in transparent bottles rather than in the traditional brown or dark green bottles used by its competitors. And we're so used to ginger ale coming in green bottles that we probably wouldn't buy it at all in a clear bottle.

Conservative use of corporate colors is usually the most effective strategy; they should function as a reassuring element, reaffirming the seriousness of the organization.

Colors for brands and products are something else again. By now everyone knows that market research shows that the most attention-grabbing color is red; but if everyone used that research as their guideline, all supermarket products would come in red packages, an obviously silly circumstance. On the other hand, straying too far from "proven" colors can be fatal. There was once a snack product called Screaming Yellow Zonkers that came in a predominantly black package. The package color deadened the product's image. The use of color in a supermarket is a delicate balance; it should be used partly to attract attention and partly to shape image.

As important as colors are typefaces that present a company's communications. How modern would Sony's electronic products seem if its logo was expressed in an old-fashioned Spencerian script? Isn't it pragmatic of Sony to use a font that appears as if it came from a computer? Sara Lee, when used as a brand identification on a package, appears in a bold, signature-style typeface—very different from the corporate Sara Lee typeface, which is a solid, stolid serif. Different audiences, different treatments.

Caution: Overwrought typography may be hazardous to corporate and product health. Good typography should encourage legibility—it should help make your name easy to read. Tortured or overdesigned type styling may satisfy a designer's artistic expression, but it ignores the reaction of target audiences. The Japanese company Anritsu, which attempts to position itself, through an advertisement in *BusinessWeek* (May 7, 1990), as "pioneers in measurement," is setting a new standard in bad typography. The design for its logo, which is a version of its name, is almost impossible to read. A typical peruser of the magazine would simply have flipped past the ad, even if he or she were interested in buying measuring equipment or investing in the company that makes it. "Tiny glitches lead to Giant Wobbles," the ad says, in a pitch for good measurement. Need I say more?

Another important factor to consider is how the product or service appears within its environment. The golden arches of McDonald's can be seen for miles, and are for obvious public consumption. Conversely, some jewelry, leather, and high-priced clothing shops along Manhattan's Fifth Avenue maintain such an understated presence that only a discreet sign on a small table announces it; this practice presumes that the customer already knows the establishment's products or services before entering the store.

Today, almost all localities have zoning requirements aimed at protecting the area's aesthetic appearance. Marketers are frequently prevented from simply putting up exterior signs that have been approved by corporate headquarters and are part of an overall design system. Signs and building styles often have to be drastically modified to conform with local regulations. Signage must anticipate limitations and inconsistencies in siting, while serving its function as an image-shaping device. Because it is so subject to outside constraints, signage is the identity tool over which companies have the least control.

The uniform or general appearance of the company's representatives is also crucial: Would you believe that you were buying your groceries at a reasonable price if the young man at the checkout register of the local A&P was wearing a tuxedo? Would you feel reassured if the pilot of your transcontinental flight showed up wearing a bathing suit, filthy T-shirt, and cowboy boots? The reaction of external audiences to a particular style of uniform is important, but even more so is how the employees themselves feel about it. If your employees find a uniform offensive because it suggests regimentation, in turn suggesting constraints on behavior, change the uniform. Design it primarily with the objective of making the employees feel good and comfortable, for uniforms can be a unifying force for the company, a symbol of belonging to something worthwhile and a source of pride to the wearer. A uniform should do more than just keep regular clothes clean or designate the rank or station of an employee. Properly designed and used, uniforms can be a good image-shaping tool, and a status symbol—but for the organization, not the individual.

Tone and Manner

Tone and manner are an all-encompassing blanket wrapped around all that is visible and audible in a company's communications; these can be powerful perception-shaping ingredients. All company communications have a tone and a manner. These qualities refer not only to what is said, but to how it is said and how it is perceived. While it is relatively easy to control and coordinate tone and manner at the corporate level, or in the marketing of a single product, the subject becomes considerably more complex when several different products or services are marketed under a single, shared identity. A Crazy Eddie can scream his messages—regardless of content—in frenetic bursts and a manic, high-pressured style. The message, the tone, and the manner effectively reinforce the desired image: cheap, cheap, cheap; fast, fast, fast. And American Airlines can focus on shaping a single image, because that image supports the only product it offers. Since American positions itself as the business traveler's airline, it will not show female flight attendants prancing around in bikinis, but it will feature smoothly delivered advertising messages, soothing and nonstartling interior plane designs, and a crisp, well-organized, and professional tone and manner. All of these elements reinforce the desired image for this single service.

Contrast American Airlines' simple task with the more complicated needs of a company that offers multiple products and services, where marketing objectives can differ from product to product, and you'll see that the very differences among the products offered and the marketing objectives pursued can cause havoc along the tone/manner dimension. For example, AT&T sells its long-distance services to both residential and business consumers.

Research has indicated that both groups of customers seek the same things from the services they buy: value, quality, and service. From an image point of view, the problem is how each of the two distinct customer types perceives value, quality, and service. To the residential customer, value means something that is easily affordable; to the businessperson, value means something that is cost-effective. To the residential customer, quality means clear communications; to the businesssperson, quality has to do with stature and the reassurance that he or she is making a safe buy. To the residential customer, service means friendly operators; to the businessperson, it means professional response to service problems.

These different interpretations of value, quality, and service mandate the company to communicate with each group in a different tone and manner. AT&T's communications to the residential market project a warm and fuzzy feel. Motherhood and apple pie. A welcome part of your home and life.

The problem for AT&T arose in 1988–89, when it tried to market to the business audience in a tone and manner that contrasted sharply with the warm, fuzzy commercials aimed at the residential audience. AT&T decided to direct a threatening message at the business group: if you don't choose AT&T, you could lose your job. Fear was the underlying theme of this campaign.

While each communication was correct for the two separate and distinct marketing plans, their conjunction created problems for AT&T; the image of the "master brand"—AT&T itself—was subjected to sharply different pulls. Was AT&T warm and friendly, or cold and menacing?

Will the real AT&T please stand up! Clearly, this complicated image problem can be resolved only if both divisions are aware of the significant impact of tone and manner on image, and formulate a plan to coordinate all messages to customers along an agreed-upon set of guidelines that accommodate the needs of both marketing plans—while protecting the long-term image of the master brand.

Caution: Coordinating messages will become increasingly difficult as the cost of communications continues to escalate and products and services continue to proliferate. Corporations will seek to leverage existing levels of awareness by extending brand and corporate identities to an ever-widening range of products, in a broader range of media. Diverse products will be marketed to diverse audiences under a shared name. To make the leveraging at all effective, tone and manner will become increasingly of the essence. A positive image will be viewed as the goose that lays the golden eggs, and marketers will increasingly come to the understanding that the goose needs as much security and tender loving care as do the golden eggs themselves.

History and Industry Segment

The industry to which a company belongs has a major impact on the way the company is perceived. The garment industry has always suffered from a dreadful image: wily business practices; a "here today, gone tomorrow" mentality; corporate instability—quintessential short-term thinking. How can individual companies separate themselves from this industry image? Liz Claiborne, though nothing more than a garment manufacturer, ran her business like a marketer. She positioned herself and her products as expressions of a life-style, using that imagery as her weapon to overwhelm the negative associations of her industry. By skillfully focusing her audiences' attention on the promise of her product, rather than on the manufacturing process, and appealing to the ultimate consumer rather than to her distribution channels, Claiborne managed to rise above the fray. On the other hand, until very recently the giant company Leslie

Fay ran itself as a manufacturer, and consequently was perceived as being typical of the industry in which it operated. Ignoring the value of the Leslie Fay image (good garments at fair prices) among its ultimate consumers, the women of America, the company relied on the trade for sales support. In doing so, it lost control of its entire marketing process. Did that matter? You bet it did. Overwhelmed by its industry image, Leslie Fay had a typical garment industry stock multiple of 7X to 10X, while Liz Claiborne enjoys a 13X to 15X multiple, continues to build equity in its identity, and is perceived as a fashion leader.

The principle is the same throughout many different industries: *separation from the industry image is critical to a company's individual image.* No matter how safety-conscious a utility may be, these days, if it generates electricity through nuclear power, it has an image problem. For many years all banks assiduously fostered a reputation for integrity, the safekeeping of money, and convenient service. Then came the enormous savings and loan crisis, in which many S&L's went bankrupt and just as many others were tarred by the failure of the bad ones. Now it has become imperative for banks to differentiate themselves from the sort of S&L's that many people have come to distrust.

Last but not least is the insurance industry, which has always suffered from a negative image—perhaps because it deals with the unpleasant issues of death (life insurance), catastrophe (property and casualty insurance), and sickness (health insurance), or perhaps because of its traditional high-pressure sales methods. Regardless of the source, this negative image has a continuing effect on individual companies within the field, requiring them to separate and distinguish themselves through their symbols. Prudential's rock signifies stability, which obscures the unpleasantness of the subject of insurance, Allstate's clasped hands send a comforting message, and Travelers' umbrella conveys well the idea of protection. These three companies project greater caring and human dimension than do companies whose symbols are their buildings, or mythological figures, or cold, verbally meaningless acronyms. *Caution: Companies that do not differentiate themselves from the industry image are compelled to share it and take the consequences.*

The Corporate Culture

A company's own corporate culture is a major influence on how it is perceived, and may prop the company up or hold it down. Corporate culture is the least well-understood component of a company's image, perhaps because the idea of culture is seen as irrelevant to business, a nicety without real impact. Interestingly, in our research we have found that corporate culture is critical, especially to a company's bottom line, though many companies have no way of assessing or understanding how it affects their financial health.

Perhaps I can persuade such companies to understand this subject by couching it in a communications context. *Corporate culture is the sum of the conversations that a company holds with itself.* It is, therefore, a reflection of the values and priorities that are held and projected by top management, and a description of what the corporation expects from its employees.

A corporate culture consists of numerous elements that relate to the employees' perceptions of the company: whether employees feel they are valued or dispensable, whether they believe the products are of high quality or shoddy, whether they feel that a job at the company projects integrity or some lesser virtue. Since all employees interact with others in the company in various ways, they become the carriers of and ambassadors to the corporate culture—and a powerful influence on the company's image. What the employees do and think within the company's walls also affects outsiders' views of the company. Employees' beliefs about their own company are invariably communicated to the outside world of customers, creditors, suppliers, and potential employees. When the troops believe that the company has their best interests at heart, they become its greatest boosters. IBM's employees are almost always the company's best external ambassadors, even when they are not used primarily as salespeople, because they feel good about their employer. Conversely, when employees feel used or abused, the company's image suffers in proportion. Drexel Burnham Lambert was just as severely damaged by its bankrupt corporate culture as by its bankrupt finances.

To best utilize this resource for shaping a company's image, management must first and foremost communicate to employees, in clear language that eschews lofty phrases and empty clichés, the values of the company—what it considers important and unimportant. In most companies, this is simply not done. Managers are left to improvise what they imagine to be acceptable corporate behavior, sometimes with dire consequences. In the 1970s ITT did not make crystal clear to its employees that large political donations and shredding of evidence about its interference in the internal political situation in Chile constituted unacceptable corporate conduct, but by its oversight reinforced the mistaken notion that such meddling was acceptable. When their actions were exposed during Watergate the image of ITT was severely damaged.

In a few companies, the setting out of such rules of conduct is embedded deeply in the corporate culture. Having worked with Goldman Sachs, I cannot conceive that the firm would ever suffer the fate of Drexel Burnham Lambert, pleading guilty to an infraction of SEC regulations that would require fines of $650 million. Employees at Goldman Sachs know what is expected of them in terms of behavior, and what is and is not acceptable conduct. Information has been regularly conveyed from the top echelon on down, in clear and certain terms, and then backed up by deeds. One of the consequences is a jewel-like corporate image.

If, after reading the company's guidelines for behavior, an employee mutters, "Well, of course—what else would they say?" then the company has failed. Rules for behavior must make sense, and not be couched in phraseology that is easy to dismiss. Also, top management must never act in such a way as to contradict fundamental precepts. Johnson & Johnson's mission statement assigns the interests of its customers—the families and doctors it serves—as its number one priority. When Johnson & Johnson's reaction to the Tylenol crisis reflected that top priority, a clear message was sent to employees that the company meant what it said about having the interests of its customers at heart. As important, management's fidelity to this commitment reinforced the credibility of its pronouncements across the board, an achievement any company's leadership would devoutly wish. *Caution: A corporate culture is only as good as management's commitment to the company's basic credo.*

Having established the elements that present a company to its audiences, and how they can be utilized, we must now examine the circumstances under which a company should consider taking a closer look at these identity elements, to determine whether they are truly serving the company's short- and long-term business plans—in other words, whether they are widening or narrowing the gap between desired image and reality.

CHAPTER 4

Symptoms, Diseases, Treatments

The need for a new or an overhauled corporate identity does not mushroom overnight and take a company by surprise; the symptoms of malaise develop for some time, and are always related to underlying business problems. An inappropriate identity is a visible symptom of a corporate disease; generally, when the signs can be readily seen, it's time for action. Sometimes CEOs, upper management, and boards of directors recognize these signs quite readily; in just as many instances, corporate officers fail to see them even when they are apparent to outsiders. Over the years, we have identified a half-dozen "red flags" that signal companies to reassess their corporate identities.

In the following pages I identify the flags and provide examples of some companies that took action when they spotted certain flags waving in the breeze.

Six Flags of Warning

Flag #1: Bargain Basement
While the company's key financial ratios and long-term profitability are extremely healthy and favorable, the company's stock price does not adequately reflect that health, and the officers of the corporation believe the stock is undervalued. Despite attempts to trumpet the company's vigor, the stock's price seems unable to reflect its "true" worth. This state of affairs is generally

linked to a whole catalog of corporate curses: a reputation for underachievement, a perhaps-unfair aura of poor management; executive stock options that are "under water," and, last but not least, a heightened vulnerability to hostile takeover bids.

A few years ago Bausch & Lomb was perceived as an unexciting, low-growth company whose primary business was in optics and contact lenses—when the company had actually reduced its reliance on optics, especially the scientific products, and had successfully diversified into a variety of high-growth health care products and services. The company of old had become something new, but not enough financial analysts either knew or appreciated the dimension of Bausch and Lomb's success and change. We recommended that the company adopt corporate identity practices to reflect its repositioning, and a year later its stock rose 52 percent; while other factors clearly contributed to the dramatic upswing in stock price, the company credits the revised corporate identity practices as a catalyst for the increase.

Flag #2: Yoo-Hoo—We're Over Here!

The officers and board have difficulty commanding the attention of the broad range of security analysts who ought to be tracking the company's business performance. Through acquisition or other changes, the company may have broken out of a narrow category in which it was formerly lumped with others in a single industry, but has been unable to convince Wall Street that it belongs in a new category, or that it should be judged differently than it was in former years.

Analysts assigned to the chemical industry were for years the only followers of Allied Chemical's performance. In the mid-1970s, the chemical industry as a whole had an extremely low price-to-earnings multiple, and Allied Chemical's stock was stuck in that multiple. If the company had been producing and distributing only chemicals, the multiple would have been unavoidable, but since CEO Ed Hennessy had diversified the company so that less than 20 percent of its revenues came from chemicals, he wanted a better multiple. Among the changes we recommended was to alter the firm's name to Allied and to drop the portion of the name that seemed to glue it to the chemical industry; within a year of adopting this and other new corporate identity practices, the multiple had improved by 50 percent. Again, we can't claim all the credit for this rebound, but without the new corporate identity practices, the company agrees, the increase would not have been as large.

Flag #3: Runt of the Litter

The industry in which your business competes is booming; nonetheless, your own company's market share is stagnant. The company is frustrated at watching the entire industry grow in size and power while it is unable to keep pace itself.

For over a hundred years the Great Atlantic & Pacific Tea Company was considered not only the pioneer among supermarkets but the premier chain. However, in the 1960s and 1970s its fortunes ebbed, its market share eroded, its management stagnated, and after a last split in 1971 the stock price declined badly. A new management took over with a mandate to restore the company's vigor and innovation, as well as its leadership position in the industry and pattern of growth it had enjoyed in the good old days. Toward that end, management instituted a host of changes that encompassed merchandising, responsiveness to the new life-styles of its customers, store design, and so on. These were important changes, but they were not enough to lift the company out of the doldrums. Management also recognized that a successful repositioning required that consumers perceive the company differently; thus, the company needed a look that said it was different than it had been in former years. A vital part of its rebirth, then, was an entirely new logo and identity system, which supported the inner changes management had instituted and wanted to communicate to consumers. If the changes had remained behind the scenes, so to speak—if a new logo, graphics, and store design had not dramatized them—the company might not have seen its receipts rise as readily at its cash registers.

Flag #4: Help Wanted

A three-month search for a new head of marketing—or finance, or distribution—has yielded few candidates, and none of them with the right background; several people whom you have courted have turned down your offers.

New York Life, one of the nation's oldest, largest, and most respected mutual life insurance companies, decided to expand its range of products to include "non-insurance" investments such as equities, mutual funds, and bonds. Among the problems the company faced was attracting competent salespeople to sell these products. Many good potential candidates felt that the image of the no-risk insurance company—a company whose products came with no risk, but also produced little gain while you were alive—hurt potential sales of investment products, even though the image undoubtedly helped convey to clients the company's financial stability. Our recommendation was to allow the salespeople to wear two hats, one that identified all non-insurance products as NYLIFE investments, and another that identified all traditional insurance products as New York Life products. Now, persuading very conservative life insurance executives to embrace any sort of change is a task that demands persistent and Herculean energy—but the logic of our recommendations was strong, and could not be ignored. It was the key to overwhelming the inbred resistance to change. When the two-hat strategy was adopted, this simple modification not only enabled longtime insurance salespeople to differentiate their product lines, but, as important, helped the company attract a new breed of investments salespeople who would never have worked for "just an insurance company."

A more humorous but no less serious problem was encountered by Tampax, Inc.: a difficulty in signing up choice recent Harvard and Stanford MBAs for its executive training programs. A significant number of these graduates felt uncomfortable about telling their family and friends that they would be using their expensive, sophisticated knowledge to service an unglamorous company whose sole output seemed to be a feminine hygiene product. There was nothing wrong with either the product or the company, all agreed, but you just didn't want to write home about it. This reluctance on the part of new recruits was one reason, albeit a minor one, that Lippincott & Margulies recommended changing the name of the parent company from Tampax, Inc. to Tambrands, Inc. While this new identity repositioned the company as a more-than-one-brand corporation, the name change yielded the additional benefit of sounding better to a very important audience of tomorrow's executives.

Flag #5: Small-Town Stuff
Your company has dreams of expanding its horizons from a local area outward, and has a good, sound marketing plan and internal resources to grow in new directions. In the past, the company has been strongly identified with a local or regional market, but now the company wishes to go beyond its previous borders. However, the financial community and/or your existing customer base greets with skepticism your expansion plans.

Study after study has shown that, for varying reasons, the financial community continues to perceive regional companies as provincial and ill-equipped to compete in the international arena. In the mid-1980s, Sohio was thought of as a limited gasoline retailer operating only in the state of Ohio. This was absurd since the company owned the largest domestic oil reserves in the United States, was a leading producer of minerals and chemicals and a leading retailer of gasoline at pumps in many states beyond Ohio, and was prominent in gas and oil exploration throughout the world. The company had large plans for expansion, and wanted to be perceived properly. So that it could overcome the skepticism of the financial community, we suggested that the company relegate the Sohio name to the brand of gasoline it marketed in Ohio, and to adopt as its corporate identity the grand old name of the Standard Oil Company, which had been the corporate name in the days of John D. Rockefeller. The name's prestige and historic identification quite quickly eclipsed the company's old regional image and helped position it as the international player it had already become. The changeover did not go as smoothly as we had hoped. The very day after the name change was announced, Chairman Al Whitehouse and President John Miller were fired. Fearing that their dismissal was—or would be—linked to the name-change decision, we really panicked. Would we ever work again? Fortunately, it was a false alarm for us, if not for Whitehouse and Miller. The entire board had supported our recommendation because we had taken the time to carefully build consensus throughout the process, and so we—and the new identity—came through unscathed. Had the identity not done so, all other company managements might have been terminally reluctant to embark on an identity overhaul program, thinking that it might also result in their current employment status being altered along with the company name.

Flag #6: Undersubscribed

Senior management gets approval from the board to issue new stock, and then is chagrined to find that it is undersubscribed. Apparently, major institutional investors do not fully appreciate how successfully your company has capitalized on the new, deregulated environment, and isn't buying what you're offering.

Management of The Union Mutual Life Insurance Company wanted to take the company public, a process that involved going from "mutual" ownership to stock ownership. If the name remained the same, there existed the danger that the company would still be perceived as mutually owned, and consequently, that the stock might not be properly bought up. There were also legal considerations—the state would not allow the company to retain *mutual* in its name if the ownership structure was altered. On the other hand, the company wanted to have continuity with the past and not lose valued customers. We recommended UNUM as a corporate name that would link the company to its traditional heritage while signaling its new and unrestricted future. Our recommendation was greeted with more than the usual hoots and hollers of derision (a hazard of the trade), because while the roots of the word were fairly easily explained, the word itself struck listeners and viewers as particularly strange and alien. This hurdle we leaped by the novel notion of asking board members to reach into their pockets and take out a quarter coin. When they got to the phrase "E pluribus unum," the naysayers knew they were beaten. Not quite wrapping oneself in the flag, but near enough.

Today, there is no doubt in the management's mind that this change was absolutely necessary, and that retention of the old identity would have compromised its strategic plans as well as its ability to raise funds in the capital market.

Okay. Those are the six flags. We've identified four further major reasons to consider a new or somewhat altered corporate identity. To differentiate these from the others, let's call them flare rockets—some of distress, others of joy, but all to call attention to something happening with the company.

Four More Rockets

Rocket #1: M & A

When your company is acquired or merged with another, or when it has acquired some new assets, an old identity may no longer be adequate to describe, reflect, and advance the new corporate reality. Under Michel Bergerac, Revlon acquired so many health care companies that more than 50 percent of its revenues derived from non-cosmetics or fragrance sales. Despite this alteration in the company's character, it failed to adopt new corporate identity practices that would have reflected its newfound diversity. Revlon continued to be perceived almost exclusively as a cosmetics company, and coincidentally began to lose market share in its historic core businesses. Its image became tarnished, its shares undervalued. This ultimately led to a hostile takeover by Ronald Perelman. While I am not suggesting that an altered corporate identity program would have saved Revlon from a takeover, had Revlon been better understood as a broad-based health care conglomerate, management might well have attracted wider support, or been able to achieve a higher price per share, thereby reducing its attractiveness as a takeover candidate.

Rocket #2: New and Improved?

When there is a major change in strategy in the company, such as divestitures that alter the alignment, say, to a "pure play" from a conglomerate, or a pricing policy that goes from mass-market to discount, the old identity may no longer apply.

Sears, Roebuck is always described as America's leading retailer, and invariably refers to itself as Sears, which is the communicative identity for its stores, though not for the parent corporation. Because this retail linkage is so deeply ingrained, the company is always judged by its performance in retailing clothes, appliances, and other items usually carried by large department stores, and is measured against other major retailers such as J. C. Penney and Wal-Mart. Recently, Sears's retail sector has been having some sales difficulties, and, accordingly, Sears's image has been adversely affected—even though the other elements in Sears, Roebuck are doing well. A new positioning would enable Sears to take advantage of the significant performances of its subsidiaries, Allstate Insurance, Coldwell Banker real estate, and the PaineWebber brokerage firm.

Further, Sears may have undervalued its own status and prestige with the buying public when it announced a new "everyday low pricing policy." From an image standpoint, this was problematic at best. Sears had earned a unique spot in American culture, and was a permanent and much admired part of the American scene, the store everybody had grown up with. America shopped at Sears because people believed that Sears delivered quality, value, trust, and integrity—always and everywhere. I can't stress highly enough that such an image is priceless; in fact, it cannot be bought, and must be earned over a period of many years. Yet Sears ran the risk of devaluing its own image and decided that discount pricing would have a stronger appeal than the image that had sustained the company throughout its history. Unfortunately, low price policy was not proprietary, since K mart, Wal-Mart, Target, and all other discounters shared it. It was also a strategy that risked reducing profit margins and that said, loudly and clearly, that all Sears had previously stood for was no longer meaningful or of sufficient value.

Then Sears confounded the problem by not altering the look of its stores or its logo or other graphic expressions of its identity to reflect this new pricing policy. Consumers saw the old bottle, but it contained new—and cheaper—wine. By looking precisely the same as it always had, but acting differently, Sears flirted with weakening what had once been a clear competitive edge.

Rocket #3: Post-Disaster Syndrome

When a company suffers a major disaster—a terrible oil spill, the discovery that cans of the company's soup have caused botulism, or identification with a discredited strategy—it may well be time for a reevaluation of, if not always for a change in, corporate identity.

Consider Union Carbide in the wake of the tragedy at Bhopal, India. The company's response to the disaster was admirable, and included having the chairman visit the site, pledge to correct the source of the problems, and pay the victims and their survivors; the negative impact of the tragedy was softened by the company's good public relations effort. But after the headlines became yesterday's news, Union Carbide continued on with precisely the same corporate identity practices as it had previously used; it continued to use them even after divesting its consumer products divisions, an action that radically altered its corporate reality. Union Carbide itself had changed profoundly, but its identity remained the same. In fact, the identity was perceived as an asset not worthy of altering even though the company had suffered through a major tragedy and a structural reorganization. Although it is impossible to say whether Union Carbide would be more highly regarded today if it had paid more attention to its corporate identity program, it is clear that management has ignored some powerful rockets, and sophisticated observers might wonder if other major assets besides the corporate image have been undervalued or misunderstood.

The case of Union Carbide shows clearly why an image program cannot simply be event-sensitive. Here are some other important instances.

When an event of the magnitude of the Alaskan oil spill occurs, all hell breaks loose. The news media have a feeding frenzy. Public attention is focused on arenas ordinarily overlooked. The government's scrutiny is awakened. When black ooze from its own tanker spread over Prince William Sound, Exxon reacted in a way that didn't help its cause. There are many explanations for the company's inadequate behavior in the crisis, but chief among them must be the low priority assigned to public affairs, as exemplified by the relatively tiny public affairs staff at Exxon corporate headquarters. Reducing the public relations apparatus had been part of a cost-cutting plan that new management put in place in the service of making Exxon lean and more profitable. The March 16, 1990, issue of the *Wall Street Journal* suggests that Exxon's troubles stem from the time of restructuring: "Industry experts as well as many Exxon insiders believe the company has a belated case of 'restructuring blues,' including a workforce stretched too thin, shaken management confidence, and sinking morale." But such cost-cutting in regard to public affairs did not exist in a vacuum; actually, Exxon had been stiff-arming the press for many years, saying that what the company did was of little concern to anyone except the stockholders, and that the stockholders' primary interest was in profits. Journalists habitually looked on Exxon's communications with suspicion. Those were the historic factors. Now to the on-the-spot difficulties. It wasn't until several days after the event that the spill was recognized as a crisis for the company. During those first, crucial days, high-level Exxon officials and spokesmen were not on the scene in Alaska, but, rather, issued communiqués from afar—an action that bespoke the company's view of the spill as less than important and that was interpreted as expressing the company's disregard for public opinion. The situation was exacerbated by Exxon's seeming unwillingness to do anything to clean up the spill unless forced by governmental action. Then Exxon gasoline prices increased at its retailers' pumps; the public concluded that the price rise was tied to the spill—and, although the several-cent increase was obviously due to other factors, this, too, adversely affected Exxon. Soon consumers were buying their gasoline elsewhere, and tens of thousands of people cut their Exxon credit cards in half and sent them to company headquarters in protest.

It didn't have to happen that way, but it did, in part because Exxon believed that image was a sometime thing instead of a long-term issue, and tried to get away cheaply by inadequately funding the public affairs function, and by refusing to understand that it takes work and effort to create a good reality and an image reflective of that reality. Today, after several more spills, most in the New York metropolitan area, the damage to Exxon's image has been severe, to the extreme of a published comment from New York City's environmental protection commissioner: "The name Exxon has become a household word for environmental irresponsibility." While clearly extreme, the fact that this position has any support is a problem in and of itself.

Only an in-place positive image can help a company weather such a crisis. Consider, in contrast to Exxon's continuing problems with the Alaskan oil imbroglio, the Tylenol poisoning scare of a few years ago. The manufacturer, Johnson & Johnson, had a long-standing reputation for 1) truth-telling; 2) communicativeness with the press; 3) aggressive, consumer-oriented actions; 4) a credo in place that set out in clear, certain terms the company's priorities, prime among them a concern for the health and welfare of the families and doctors served by the company. These attributes stood the company well when bottles containing Tylenol were found to be the cause of several deaths. Rumors spread, press inquiries flooded in, and there was a danger that the impact could not be confined to the sales of Tylenol but might (you should pardon the expression) spill over and adversely affect the sales of other Johnson & Johnson products.

But what actually happened? Following the guidelines of its credo, the manufacturer took control of the crisis by getting out in front and handling it from headquarters, and by issuing reams of factual, up-to-date information on the recall of all Tylenol products and, later, on the new measures being taken to sell the medication only in tamper-proof containers, well in advance of new government regulations for such containers. Tylenol was effectively distanced from other J&J products, so that the sales of the others were not affected, and within months, Tylenol was back on the shelves of drugstores and supermarkets and the crisis was over. It's old history now, not a festering sore that remains open despite efforts to close it, as is the case with the Exxon spill.

Now some people argue that all that is needed to deal with a crisis is a contingency plan. In a published statement, Exxon CEO Frank Rawls said that if Exxon had had such a plan, the oil spill's effect on the company would have been reduced. Let's consider crisis management for a moment. It seems like a good idea, but are contingency plans enough to deal with crises like oil spills and tainted consumer products? Can upper management really take the time from its other duties to formulate adequate contingency plans? Is it possible to game out every possible crisis, and make a plan to fit it? In actuality, no crisis ever fits precisely into the parameters previously gamed out. Since it's virtually impossible to imagine in detail every potential nightmare, the real requirement is to have an overall plan for dealing with crises, to establish the context in which your later actions may be judged. This, in turn, mandates an attitude that both in good times and in bad, image matters. A contingency plan to clean up the spill, as well as to handle public reaction, certainly would have helped Exxon, and a similar plan did help Johnson & Johnson—but Johnson & Johnson's evaluation of the significance of image, its willingness to invest in its nourishment, and its commitment to making the necessary resources available to do the job were the true differentiators. Having managed its image extremely well for many years, the company was able to deal with its crisis from a position of true strength. Its positive, honest, forthcoming image allowed the employees and divisions to rally to the company's side, furthered communications with the press, and afforded the company time to make the necessary adjustments in the product packaging without losing too much ground.

Another recent example of a company reacting well to a crisis was AT&T's splendid effort following January 15, 1990, when, for the first time, AT&T long-distance lines simply broke down. The impact of such a failure on business, on individuals, even on our national defense could not be exaggerated. The crisis was obviously unanticipated, but AT&T reacted to it very well— in my view, because it has long attached a high priority to the maintenance of its image in the eye of the public. AT&T first attempted to understand and fix the problem. Second, it kept the public informed in a forthright and intelligent manner about the difficulties and their resolution. Third, after the repairs were completed, Robert Allen, Chairman of AT&T, let it be known that AT&T would file with the FCC a request to offer substantial discounts to all its long-distance customers for one day, to make up for the disruption in service; to cap it off, the day chosen was Valentine's Day, a time of traditionally heavy long-distance lines usage. A class act, not only charming and intelligent, but the right thing to do, and a solution that made good business sense as well. A survey of 193 corporate telecommunications managers conducted by Business Research Group of Newton, Massachusetts, in March 1990 showed that of the companies affected by the disruption, three-quarters of the respondents said they were satisfied with the company's response, and nearly all said they believed the temporary outage to have been an isolated event. AT&T was decidedly happy that 89 percent of these crucial customers had no plans to consider switching to MCI or Sprint. The only disappointment for AT&T was that less than half believed that the company's Valentine's Day charge reductions were an adequate way of compensating for the problem.

Both AT&T and J&J had spent a lot of money over a period of many years to produce images that helped them during a crisis by *predisposing the public and the press to believe that what the companies were doing was to the point and in the public interest as well as in their own interests.* Exxon had focused its image-shaping activities solely on brand marketing. It had strong communications plans only insofar as they related to the company's brand of gasoline, and it almost ignored the significance of corporate image. Frankly, had the corporation not marketed a brand that shared a name with the corporate entity, I suspect that there would have been less effort to clear the Exxon name through spill-clearing activities. In 1990, with spills off Arthur Kill in New York harbor, Exxon repeated the image-related mistakes it made after the Alaska spill. Despite the fact that Exxon in recent years has achieved an exemplary record of profitability and stability—for which it is admired within the business community—in its bungling efforts in image-related matters it seems to exemplify the old adage, They know the cost of everything but the value of nothing.

Rocket #4: Changing Trends

Crises are not always caused by events. Difficulties can come to a head as the result of changing trends. Coca-Cola's reaction to such a crisis/trend is a case in point. Recently, the Coca-Cola Company announced the introduction of Coke II to its line of products. Coke II was described as a renaming of its earlier new product, New Coke, whose introduction in the mid-1980s had proved disastrous but whose sweeter formula was being brought back again because, time after time, in carefully supervised blind taste-tests, the sweeter taste consistently won. New Coke, and now Coke II, were reactions to the inroads made in market share by other colas that were sweeter than Coca-Cola.

To my mind, the thinking behind the strategy is only half-right. Each new generation of soda drinkers needs to have its own particular product; the most successful effort in this regard was the "Pepsi Generation" advertising campaign. In that instance, the product, package, and everything else about the product was not changed, but the advertising was dramatically altered, and so, eventually, was the image of the product. With Coke II, Coca-Cola is trying to hook a new generation by means of a change in the product itself. Recognizing the psychological requirement of interesting a new generation was the half of the equation that Coca-Cola has gotten right. But it is ignoring the lessons of its own industry's past.

In the early 1970s Pepsi Cola, having failed with a diet soda called Patio, launched Diet Pepsi, which became an immediate success. Coca-Cola, recognizing the potential of a diet cola, nevertheless refused to permit any product other than its original formula to be identified as Coca-Cola, and called its own diet cola Tab. Coca-Cola as a company steadfastly stuck to the policy of having only the original formula identified as Coca-Cola until future projections showed clearly that the soda market was going to be dominated by soft drinks. Should that happen, it might be in danger of becoming known as the Tab Company. To avoid that, it reversed its original policy, and launched Diet Coke, which also became a great success. I admired the company's turnabout in policy, since it was an attempt to keep up with the times, which is always necessary in business.

But with the extension of its line into Cherry Coke, Coke without caffeine, Classic Coke, and, now, Coke II, the company has encountered a new problem. With so many products, it risks confusing its customers and disappointing either old or new Coke drinkers who mix up the product or the name. For instance, distribution to supermarkets, fast-food outlets, and vending machines is complicated by having to stock Coca-Cola and Coke II at all locations. The longer list of products will also force customers to spend a lot of time and energy considering the subject of Coca-Cola—and this, clearly, is a mistake. A primary objective of good image management is to clarify a product's image and to sharply differentiate it from all others, creating a unique and distinctive personality for each brand. Soft drinks are impulse items. But with the new line of products, we have moved from simplicity to complexity. In the good old days, a customer could say, "Give me a Coke, please." Now the customer might say, "Give me a Coke . . . or do I mean a Coke II . . . or Diet Coke, Cherry Coke, Classic Coke, or Coke without caffeine?" Contemplating all these alternatives, a customer might well be tempted to conclude, "The heck with it, I'll have a glass of water."

In short, the line extension of Coca-Cola, in its version called Coke II, may be going a bit too far. The nuances of taste that differentiate the six products of the line that all carry the moniker of Coke are of more interest to the manufacturer than they may be to customers. Management may study nuances of taste with ferocious concentration, but customers often ignore them. This brings to mind another maxim: *While developing the optimal image-shaping practices for your brands and company, every once in a while put yourself in your customers' shoes.* If you ask them to think too carefully about your product—especially if that product is an impulse item—you may be well on your way to losing them to the competition. The more complicated the message your image must convey, the more fragmented the image may become, and the greater the likelihood that the integrity of the brand will be diluted. Brand Chevrolet can be identified with eight separate models, and Brand Ford with nine, each model possessing different options and at different price points—but Brand Coke? Among Coca-Cola's many appeals is that it tastes good, is conveniently available, and makes you feel good while drinking it. It should remain the pause that refreshes, not become the pause that exhausts.

Once a warning has been recognized, how can a company resolve its business problems/image problems? In the preceding pages I've given a few examples of what can happen when companies aggressively tackle their identity problems—and some examples of what can happen if the problems are ignored. But is a corporate identity overhaul the only solution? To be sure, other "fixes" are often tried, and are *sometimes* successful.

Partial Solutions

Rain, Rain, Go Away

To be entirely honest, there are situations when one can ignore an identity problem and hope it goes away of its own accord, or corrects itself, with the company's audiences catching on that this is happening. In a recent Gallup Poll, the insurance industry is apparently home to the second least-admired group of salespeople in America—used-car dealers are in last place— but the insurance industry continues to do nothing about this public perception problem. Since for the past hundred years the industry has been, on the whole, enormously successful, the insurance companies may be justified in the hopes that the perception of their sales force does not matter.

"Let's Run an Ad. . . ."

The solution that is usually thought of to alter an image is advertising. People sit around a conference table discussing the problem, and someone says, "Let's run an ad. . . ." I don't disdain ad campaigns, for they can be enormously effective. In my view, advertising campaigns can address problems of corporate mis-identity best if the campaigns are both narrowly focused and of short duration. Consider the case of Dow Chemical U.S.A. Headquartered in the out-of-the-way city of Midland, Michigan, for many years Dow was considered quite provincial and inaccessible to the major media. During the years of the Vietnam War, Dow was attacked regularly for manufacturing both Agent Orange and napalm, as well as for being extremely insensitive to environmental concerns. Executive recruiting became more difficult, and many in-place executives began to feel embarrassed at being identified with the company, especially when they left the insular confines of Midland. For a long time, the company did not try to rectify the public's perception of it—until a new and more sensitive management took over. Dow then funded and aired a spectacular series of commercials on television in which the company was depicted as making helpful products, and as seeking out college students about to graduate who would ecstatically call home to tell their parents that they were going to join Dow because it offered them the opportunity to help humanity. The campaign did not change Dow's reality—it continues many of the same lines of business—but it had a major impact on the public's perception of the company. Certainly, no one is embarrassed any longer to say they work for Dow.

The major drawback of advertising is that it is quite expensive, and requires constant new funding to remain effective. Most executives know that, but what they don't seem to realize is that when the objective is good image management, there are alternatives to high-budget advertising. In a recent article in *Forbes* (January 8, 1990), the CEO of the company that produces Filofax laments the company's historic casual attitude toward proprietary image building, and tells how, in the late 1980s, as its signature product line began to lose market share to well-funded and cheaper competitors, the only solution seemed to be a lot of advertising. It was then that Filofax bought relatively costly space in magazines such as *Gentleman's Quarterly,* the *New York Times Magazine,* and *M,* but still lost $886,000 on revenues of $7 million in the first half of 1989. Was assigning to advertising a large portion of the available communications dollars the only route for the company to take to rebuild its image and its profitability?

Here are five notions about what else Filofax might have done:

1) Maintain the uniqueness of the product and the proprietary nature of the concept through superior design, presentation, and materials. A product that is visibly superior to all imitations can command a higher price.

2) Carefully control distribution outlets so that only upscale retailers would be permitted to sell the products. The temptation to mass-market a product is always there, but not every product ought to be mass-marketed. Frequently, certain retail images can enhance one's own brand's status. Kept in only the best stores, Filofax might have prospered greatly in a time of emphasis on exclusivity and luxury.

3) Seek a joint venture wherein the Filofax concept could be piggybacked on other well-established images, e.g., The Polo Filofax.

4) Create an identity system to enable the marketing of different Filofax models to particularized and defined segments of the market: the working woman's Filofax, the college student's Filofax, the traveling salesman's Filofax—not just the yuppie Filofax.

5) Imaginatively communicate the company's message through a variety of media other than print advertising, e.g., send a free Filofax to every secretary to a Fortune 500 company chairman.

In other words, try things that have as their basis the view that the image is a proprietary asset that demands to be managed creatively and broadly, and not simply turned over to advertising executives who believe that commissionable advertising is the only solution to an image problem. But what if your company doesn't want to "go all the way" to a new or rebuilt corporate image?

New Name and/or Logo

Many companies attack problems of corporate identity by changing their name, or by altering the company's logo, in the hopes that this will change public perceptions. The reasoning seems to be that because the company is visibly doing *something* different, the company will be perceived as fundamentally changed. This strategy often is a misreading of the company's audiences. For instance, it is unlikely that the sophisticated financial community will believe that a name or a logo change alone will reflect a new corporate reality; rather, it will be perceived as cosmetic, therefore suspect, and probably counterproductive.

Consider Bradford Trust. Here was a company that had reliably performed services for the financial community for many years, such as holding bonds or stock certificates for transfer between banks—a service whose watchword just had to be "trustworthy." For decades, Bradford had been worthy of trust, had prospered and grown accordingly, and was considered a banker's banker. Then management deteriorated and a few became involved in scandals that besmirched the company's reputation. A new management team took over and proclaimed itself determined to clean house; at the same time, behind closed doors, it questioned whether its old image had become so tarnished that it would burden the company as it moved into another era. The long-term history of Bradford was good, but the recent short term was bad, really bad. When Lippincott & Margulies was brought in to advise, we surveyed the financial community for its perception of Bradford, and were told that it had become untrustworthy. To get rid of the baggage of untrustworthiness, we recommended a radical change in communication, including a name change. Since the corporation planned to continue to provide financial data and other related services to the financial community, we created and recommended the name Fidata Corporation (proprietary, descriptive, and apt) as the foundation of this revised identity. But on the very day that the company chose to announce this new identity at a well-attended press conference, another scandal broke, this one having to do with bond embezzlement. The result was that the value of the new identity was immediately compromised. It was perceived as misleading trickery, and no amount of image making could rescue it. Management had not cleaned house well enough, and so a new identity was of no use. Later, the firm went bankrupt.

Changing just a name, or a logo, is a generally ineffective way of altering public perception, unless such a change is accompanied by other elements of a corporate identity program—a program that accurately reflects the underlying corporate reality.

Why Does Image Making Have a Bad Image?

Before going any further, and staying with the "doctor" analogy, I want to address the problem of quackery—in the context of image making, the notion that all image making does is distract the attention of a company's audiences from its realities, that it is made of smoke and mirrors. In a recent interview with *Fortune,* referred to earlier, Exxon CEO Frank Rawls opines that the crisis occasioned by the running aground of the *Exxon Valdez* tanker in Prince William Sound would have been less severe had Exxon had in place a better public relations program.

Rawls's comment reflects one of the prejudices of American business, that there is a difference between a company's image and its actual performance. Imagery is seen as a tactical weapon, deceptive or fuzzy, which can be used to conceal inadequate performance. Many American businesspeople still feel that if they do need an image, they can call and have it delivered from the take-out department of the image store. Further, they believe they only need one to hide bad performance; if their performance is satisfactory, they don't need to waste time thinking about their image.

The news media, echoing these outmoded business prejudices, often portray image as something used to trivialize a misdeed. An expert in what *Forbes* (June 25, 1990) calls "sports disaster public relations" has been hired to refurbish the image of baseball player Pete Rose after his conviction on tax evasion charges stemming from illegal gambling. The key to the "rehabilitation" strategy is "to make Pete Rose a small statistic in a larger social problem, compulsive gambling. Making him appear a victim, not a villain." Such an attempt to deny or distort the wrongdoing continues to reflect badly on the sort of positive image activities that go on in business. Another national magazine labeled the then–RJR/Nabisco CEO Ross Johnson as having a "greedy image," instead of coming right out and saying that Johnson was indeed a greedy person. It was as if the magazine presumed his image was an icon made for public consumption, but that underneath that protective covering, Johnson might well be a selfless, sharing, and generous guy. More usual are the instances in which journalists savage a warm-sweet-caring image by trumpeting that it hides an ugly reality beneath the surface; the implication is that the image is employed solely to distract attention from the reality. Both these positive and negative usages of the term *image* are attempts to find a difference between the image and the reality, and to divorce image from reality.

As a consequence of believing in the separation of the image from the underlying reality, many American businesses continue to think of image matters only in relationship to crises. And they confuse awareness of the company with having a good image. To succeed, they presume, all that is necessary is to be well known; consequently, they focus most of their image-making efforts on raising levels of awareness of their company, often to the exclusion of all other factors. Put in other terms, they concentrate on the tactical aspect of image management, and make every move in lockstep with their public relations and advertising campaigns.

The problem comes into clearer focus as we look at some of the meteoric companies of the 1980s. Many of them were, or became, very well known, and used the media as the arena in which to make their grand plays. Drexel Burnham Lambert, Saatchi & Saatchi, Shearson Lehman Hutton, and Donald

Trump are just a few of the casualties of the 1980s whose misunderstanding of image management contributed to their downfall.

In his most recent book, Donald Trump concedes that his patina of invincibility has been blemished by recent disasters and says, "Look, I don't expect anyone to feel sorry for me, but the fact is I'm only human. . . . Image means a great deal to me. If people don't associate my name with quality and success, I've got serious problems. So do the thousands of people who work for me and depend for their livelihood on my doing well. Unfortunately, years of relentless striving for perfection go into creating an image, but just a few potshots from some jerks with word processors can tarnish a reputation." Using the press, as did Trump and these several companies, to shape a favorable public perception is indeed crucial to image management, but they all overlooked (or never knew) about other salient factors. For example, the total context in which you are viewed, what you say, and how close you are to delivering on your promises are equally crucial factors in the establishment and maintenance of a good image, but these factors are often ignored in the rush to become well known.

One of the associated problems of being well known is that when you are visible, your fall cannot be dignified or obscure; rather, it is rendered all the more public, and the high awareness level that once helped now serves to accelerate the downfall. Another result is that the business community as a whole begins to link high awareness with failure, thereby damning all image-making activities. Awareness is the entry criterion for business success, but it is not an end in and of itself, only a means to an end. The actual and desired end is continued good business.

Low-profile companies are, by definition, little known. Often, when a successful one is discovered, people say of it, "See, they don't talk about themselves, they just go about their business and quietly succeed." While modesty is a universally admired trait, it is not necessarily a good business strategy. In this era of multiple competitors and ubiquitous advertising, hiding your light under a bushel basket is certainly not the best prescription for commercial success. In fact, for every previously anonymous and successful company that comes to public attention there are thousands that go nowhere, and very quietly, too. In fact, in today's economy, invisibility is the exception to the rule for success. While companies don't have to go to the extreme personified by Donald Trump in his enterprises, neither should they adopt the reclusiveness of the W. W. Grainger Company—a very successful entity that I bet you've never heard of—and expect to gain a competitive advantage by remaining hidden or obscure.

The problems are that 1) the tools for image shaping are the same, whether used for honest or deceitful purposes, and 2) most audiences share the view of the deceitful communicators—that recognition is an end in itself. Because of these factors, many skilled practitioners are able to use the tools of image making for misleading purposes. Many audiences have no interest in the story being presented to them, have no urge to learn whether there is a disparity between the reality and the image. Corporate managers habitually underestimate the indifference of the outside world to their companies, and they overestimate the attention that the public is prepared to concentrate on the companies' concerns. Once an identity has registered itself in the consciousness of the public, they have no further curiosity about it. This is the point at which manipulation can usually take place, sometimes for good, sometimes not. Bad endings to good starts can sometimes be written when victims are lulled into a sense of confidence; having heard of the enterprise, they look no further into it, and can make investments and purchases without knowing what they're buying.

Some years ago, Integrated Resources hired us to reposition it with the financial community. To better comprehend its business, we had to interview a number of its sales executives. At that time, its stock was hot—and it was in the business of marketing tax-sheltered investments to wealthy individuals. Its sales executives told us that since their prospects had all heard of Integrated Resources, and they knew it to be a high flier, it was easy to get the prospects to invest some of their money in Integrated Resources vehicles. Awareness level had opened the door. Had they looked further, the would-be investors might have seen that Integrated Resources as a company was built on a card-house of debt; not too many years after it was a high flier, the company went bankrupt. Customers were badly burned. Had we conducted some post-bankruptcy research on them, I am sure many of the customers would have excused their error by saying, "Well, Integrated was a big company. Everyone had heard of them."

An associated problem arises from the general public's misunderstanding of what is involved in changing an image. A recent article in the *New York Times,* headlined "Remaking Kidder's Image," discussed the way in which Kidder, Peabody's new owner, GE, is transforming the company, remaking the reality by changing around nearly every aspect of the acquisition. That's not an image being remade, it's a reality being remade.

I'm not just picking semantic bones with the *Times* here, but rather using this example to point up that people continue to believe a company's image is more readily moldable than its reality. They feel somehow that it's easier to take risks with the image, easier to conceptualize and articulate a change in image than a change in underlying reality. If, as I have said repeatedly in this book, the image must follow and reflect the reality, it ought to follow that image changes must not be attempted unless real alterations to the company are in progress—but this isn't precisely so. The reason is that the idea of changes in image is less threatening to executives than changes in the way the business is actually run.

Imagine a company with some difficult, even fundamental, problems that are deeply ingrained in the way it has been doing business. Its bad habits have developed over a period of years—and you know how hard it is to break a habit. To address these problems or habits is essential, yet difficult. But if these can be approached in a communications context, the concept change is invariably more culturally acceptable. Executives seem to reason, "We're not a bad company, but we're perceived in the wrong way, and we need to send out the proper messages if we're to be perceived properly." When this rationale is used, many executives tend to rally behind it, and—wonder of wonders—to simultaneously start the more difficult process of changing the way the company does business. Image changing, then, sometimes functions as a catalyst that inspires sorely needed alterations in operating procedures. Unfortunately, this role for image making is never, never acknowledged.

A third factor contributing to the bad name of image making is the relative ease with which some images can temporarily be changed, even when underlying realities are not substantially altered. Impatient entrepreneurs and hungry image-makers view the discipline as a shortcut to success. A recent example makes the point in a particularly vivid way. Barry Minkow and his ZZZZ Best Company were a classic scam. A born promoter, Minkow figured out that a story that was sexy enough to capture the attention of the media could be a strong horse to ride to wealth. By relentlessly depicting himself as a modern-day Horatio Alger, he became known, and encouraged the media to spread the word of the young man who had started a carpet cleaning company and made it a household name.

But the reality was not there. All Minkow had done was make up a story and get it repeated many times. He buttressed this publicity maneuver by associating himself with the integrity and trustworthy reputation of his accounting firm and his law firm. The combination of high visibility and the seeming endorsement of well-known and respected professional institutions was sufficient to get the public to believe in his enterprise and to convince many people and institutions to invest in it. Beneath Minkow's story lay a tissue of lies. Once revealed, it fetched Minkow a twenty-five-year prison sentence for defrauding banks and investors of about $40 million. Unfortunately, Minkow was not the only culprit of the scheme. Many people blamed his image-making activities, which were inherently deceitful, and one more lie about image making was spread.

Sometimes the quest for fame can become confused with the quest for business success, and if all that survives a subsequent crash is the fame, image making often gets the blame for the other aspect, too, the business failure. William F. Farley combined ambition and a small talent to propel himself to a modest level of business prominence. Then he decided to become famous, and to use his business funds to finance his rise in the public eye. Through high leverage he acquired some small and then some larger concerns, the largest of which was Fruit of the Loom, a company that usually spent a decent amount of money advertising its products to consumers. Farley redirected those advertising efforts to promote himself, rather than the products that the company manufactured. He became somewhat visible and known to the public, enough so to entertain the notion that he might run for president of the United States, since enough people knew who he was. When the public proved uninterested in him as a candidate for the White House, primarily because of dislike of his life-style, he turned his energy to taking over the venerable West Point Pepperell Company, a classy, well-organized textile company that held a premier position in the "bed and bath" field. Since fame, and not business success, was the name of Farley's game, he managed to raise enough money to overpay for West Point Pepperell by a full $200 million. The unsoundness of the financial foundation doomed the commercial aspect of this enterprise from its inception, but Farley's preoccupation with his own image contributed greatly to its downfall and was blamed by many observers for the fiasco in its entirety. But this, too, was not an image-making fiasco, even though the rather bad attempts at image making took their lumps because of it.

If business knew more about image making and what goes into it, businesspeople would be less likely to be duped by deceitful practices and practitioners. They would recognize the difference between a fever and the thermometer that records and communicates the fact of that fever to an outside observer. It is useless to curse the thermometer for the ravages caused by the fever; it is useful to treat well the symptoms and causes of the underlying illness. In recent months, rather than describing its effort to reshape Kidder, Peabody's image, GE has announced it is pumping nearly a billion dollars into restoring Kidder to health. This is the correct sequence of events: first, fix the reality, and then adopt appropriate practices to signal to critical audiences that they must correctly recognize the new reality.

Who Ya Gonna Call?

If your body ails, you go to a doctor, who investigates your body and eventually tells you what the cause of the complaint is and whether what you've got is serious or not, and, if it is serious, prescribes a course of treatment. If you had an unknown ailment, you would not proceed directly to a surgeon, because, among other reasons, you know that the surgeon is likely to prescribe one particular treatment for the ailment—surgery—when such radical treatment may not be necessary or appropriate.

Similarly, the first step in considering a true corporate identity program is to get your problem diagnosed, and not by just anyone. You need a diagnostician who has little vested interest in a particular solution to the problem. A good corporate identity consultant provides diagnostic services that include a recommendation for attacking the problem and a recommended range of "treatments" that may be quite startling in its extent. Here are just a few: developing image objectives; establishing coordinated communications strategies; altering the company's signage, its logo, its corporate communications, and/or its name; refurbishing campaigns to change the image of a division, or perhaps only the image of the CEO; setting up an investor relations department within the company to communicate with Wall Street; developing an advertising campaign; modifying communications architecture so that over time it will shape a new image for the company as a whole. There may be any number of recommendations—but these are for the next level of services, and should be considered only after the company has bought an objective, dispassionate diagnosis.

Is It a "Good" Diagnosis?

How can upper management or a board of directors know whether the diagnosis of its company's corporate identity problems is a good one? Of course this is always a matter of judgment, but we've found that there are several criteria that characterize the best diagnoses.

1) You know that the analysis is correct because it says insightful things about your company. The identification of the problems is not necessarily a comforting thing to hear, but it is on target and believable. Observations are not merely the opinions of the consultants, but are backed up by interviews with others. Many times, the diagnostic investigation uncovers subjects or important nuances that upper management, the CEO, or the board hadn't known about, but which are nonetheless accurate readings of internal or external difficulties.

2) The report presents these problems in an intelligent, straightforward manner, one that makes the company's executives and board members think about the company's corporate image from perspectives that they have never previously considered.

3) The recommendations of the diagnosticians, while they may be startling, are structured so that they will be culturally acceptable to the company. The solutions suggested are not so farfetched as to make the executive corps feel they can't live with the recommendations. Only those solutions that have a good chance of being taken seriously, and around which a consensus can be built, are worthy of being suggested in the diagnosticians' report. Regardless of the details of the actual recommendations, the program is designed to achieve consensus.

4) The recommendations are clearly supportive of the company's future strategic plans.

To Tell the Truth

Let's take a moment to consider an oft-raised objection to the investigations that lead toward changed corporate identity practices. Couldn't the CEO of the company, or his or her hired executives, or the members of the board of directors, find out enough about the problems of the company themselves, and determine what corporate identity practices to adopt or alter? Isn't this all just common sense?

One hopes that there's plenty of common sense in an investigation into a corporate identity, and in the recommendations made. The point is, rather, that corporate identity consultants—outsiders to the company—can ask questions and find out information that is often unavailable to insiders, and in particular is unavailable to the CEO. That's because independent corporate identity consultants are unlike most business consultants. Frequently, consultants are hired by companies to answer questions whose solutions are already known, or to deal with already familiar territory. These consultants invariably produce reports that bolster the commissioning executive's own notion of what should be done about a problem. Not so with corporate identity consultants. We are most often hired because the insiders are truly at a loss as to what solution to offer for a particularly vexing problem that is outside the usual arenas in which they perform their everyday tasks. Because of the emotional nature of the subject, and the delicacy with which it must be approached, many CEOs are relieved (if not always delighted) to be led by us, rather than to be required to lead the investigations themselves.

Often, the CEO and other top executives cannot take the time from their daily duties to investigate thoroughly the matter of corporate identity. The subject is, by its nature, subjective and emotional, and therefore more difficult than most subjects to focus on and to imagine solutions for. Moreover, certain aspects of corporate identity—the relationship between management and employees, for example—are things a lower-rank manager doesn't discuss too easily with top management.

One of the sad realities about being CEO of a large company is that you are insulated from learning many things, because your employees won't want to tell you bad news. This is especially true in regard to corporate identity, and particularly unfortunate. For problems of corporate identity *always* go to the heart of what a business is and does. Investigating these problems may require an executive to confront unpleasant material: the findings may be damaging to that executive's self-interest, or they may reflect badly on his or her subordinates or superiors; at the very least, they have a good chance of rocking the corporate boat or opening a battleground on which the executive may have neither the capacity nor the will to fight. When such an investigation is likely to bring up a lot of emotional dynamite, my belief is that it's always better for the investigation to be carried forward, and recommendations fashioned, by an outside consultant rather than by insiders to the corporation, who may have other agendas to pursue than the matter at hand. And you don't want just any kind of consultant making the investigation. Skillful and experienced identity consultants can deal with a range of corporate problems and develop the various recommended treatments in terms of communications policies—an approach that carries with it fewer intrinsic hazards than many other ways of attacking corporate problems.

Consider the question, "What business are you in?" If it were asked by an auditor who is in the process of assessing your tax liabilities, it could be viewed as threatening. If it were put to you by a lawyer who might be considering what divisions of the company ought to stay in line or go on the block, it could be more threatening. And if the question emerged from the mouth of a time-motion study man whom everyone knows is there to determine if you can get along with 30 percent fewer employees? Or from the management consultant who is usually (often, rightly) suspected of fostering a reorganization that will eventually ease the interviewee out of his or her job? Yet when a *communications* consultant poses the same question—as we invariably do—the query is less threatening, because it is offered only in a context of communications, which isn't very threatening at all. (Sticks and stones—and time/motion studies?—can break my bones, but words can never hurt me.)

Having its corporate identity reevaluated, and perhaps changed, is almost always a painful process for a company, though at the end of the journey our clients invariably tell us that it has been a salutary adventure. The investigation often reveals inadequacies, or at the very least disparities between the image the company hopes it has and the one that it actually has (as viewed by its various internal and external audiences).

The news that we convey is not always bad. The process can also reveal some hidden strengths that companies haven't known they possessed. Our diagnosis of Goldman Sachs in the 1970s, for instance, revealed that the investment "stars" of the firm were stars mainly by virtue of their corporate association and by the resources that the company provided. They developed a profound awareness of the value of Goldman Sachs the institution; subsequently, we and the firm developed both communications and operational policies single-mindedly aimed at building the notion of the strength of the institution as one of its own major assets. This posture is in sharp contrast to many of Goldman Sachs's investment banker competitors, such as the old Lehman Brothers, Wasserstein/Perella, and even Salomon Brothers. The wisdom of these Goldman Sachs policies is evident from both the strength of the company's image and its prosperity, through both good times and bad on Wall Street. Some indicators: the duration and quality of the firm's investment banking relationships; the lowest employee turnover rate on

Wall Street; and (when introducing a new investment product) the credibility afforded by the firm's slogan, "Who can you trust?"

A company must recognize that any steps taken to correct an image will have the drawback of being highly visible, and will be viewed as fair game for public comment—sometimes even ridicule. As consultants, we recognize that we are a handy factor to blame if some of that visibility causes problems. One last, wry example of having an identity consultant to kick around: In 1989, the Toyota Company introduced its new luxury car, Lexus. After having announced the name of the car with much pride and at great expense, Toyota was then hit with a suit by the Mead Corporation, which owned a legal-information service named Lexus and wanted to prevent Toyota from using the name. Among Toyota's public defenses was that the name Lexus, although created by Toyota, had been "inspired" by a name recommended by the well-known name-generating firm of Lippincott & Margulies. That was only partially true. Toyota had hired us to come up with a name for the new car, but had rejected every one of the hundreds of names we suggested, including Alexis. However, when the suit hit the financial pages, there we were, being asked to shoulder some of the public blame for the name Lexus, which we hadn't even recommended. A lower court found in Mead's favor, but a higher court has since allowed Toyota to use the name Lexus. Did L&M win or lose? Hard to say . . .

To sum up: The forging of a new or refurbished corporate identity is a seminal event in a corporation's history, one that happens usually only once every twenty years. It can provide a rallying point, a standard to which the management and employees can adhere, and in which they can take renewed pride. A company should think carefully, and act decisively, to get the most out of the difficult process of forging a new or invigorated identity.

PART 2

Successes and Disappointments

CHAPTER 5

No Image, No Premium

Many businesspeople, while granting that an image might be a good thing to have around the house, cannot exactly say what a good image might do for them. Fortunately, we can be more precise. In a recent interview for our *Sense* magazine, Sandy Weill told us, "A good brand image can remove something from just being a commodity-type product with very narrow margins to the kind of product for which you can get premium prices." Sanford I. Weill is chairman, president, and CEO of Primerica Corporation, and I have been so struck by the simplicity, directness, and accuracy of what he said in that one sentence that I am tempted to call it Weill's Law. I wish I'd said it first, but perhaps it's fitting that it comes from the man who built Shearson from a small brokerage house to a large one primarily by refusing to cut brokerage fees, and by selling service and quality to customers at a time when many other brokerage houses believed they must cut their prices to stay in business—in other words, he built his business on selling a "premium" service rather than a "commodity" sort of service that was no different from that offered by many other brokerage houses.

What Is a Commodity? What Is a Premium?

When the product or service that you're selling has become a commodity, the purchaser does not differentiate among competing suppliers of that product or service. The potential purchaser perceives no difference among suppliers in terms of 1) quality, 2) level of customer service, or 3) proprietary aspects of the product. Consequently, the supplier who will sell you the product at the lowest price is the one that carries the day. And that product becomes a commodity. The price the supplier receives for it is a direct function of its costs—and no more. Pork bellies, wheat, oil, and gold are commodities, regardless of their cost.

Conversely, when a product can command a price that is more than a simple multiple of its costs, you are in the realm of a premium. The premium is the differential between the lowest affordable selling price and what the supplier can actually obtain for his or her product or service in the marketplace. A premium price means the top price for a product or service within its own category. It does not necessarily characterize the most expensive item in the store.

If one accepts the proposition that the purpose of a business enterprise is to sustain satisfactory profit margins for the company in the long run, then it is axiomatic that the margins are heavily influenced by the size of the premium that the company is able to obtain for its products. And if one accepts that, one also accepts the necessary place of image in business, because image and premium are completely interrelated. Premium is shaped by image, and, in turn, premium shapes images.

Aspects of Image in a Premium

1) *Awareness/recognition level.* If a product or a service is not known, it cannot obtain a premium; conversely, if it is known, it can. Whether you're the best butcher in the neighborhood or you're Rolls-Royce, the principle is still the same. Raising awareness and recognition levels is fundamental to establishing an image.

2) *Specificity.* Being known is important, of course, but as important is to be well thought of. If you are known only as something notorious, awareness level may not help you. Everyone is aware of cigarettes; these days, nearly as many people are aware that they can cause a whole spectrum of health problems.

3) *Standing for something.* It might be for quality, or convenience, or reliability; whichever, this "something" is really the basis for a company's or a product's premium. Perrier obtains a premium for its products over what we'll pay for the store brand of bubbling water. We pay more for a standard item in a 7–11 than we do in a supermarket, because we're paying the convenience premium. We are asked to pay more for Michelin tires on the basis of their touted reliability.

4) *Margin of error.* Since the price obtained for a premium product is not absolutely tied to the costs of doing business, having a premium creates a larger reserve against a possible error by the company. It can experiment more and try and fail at a new thing without suffering the same consequences as a company with slim profit margins. New Coke didn't sink The Coca-Cola Company, and Edsel didn't sink The Ford Motor Company. But Crazy Eddie, whose concept was based on selling appliances at the lowest possible prices, prices that returned only a very low profit margin for the company—Crazy Eddie bit the dust when a few things began to go wrong. The larger the premium, the more room for the business to maneuver.

5) *Market share.* There is an important interrelationship among image, premium, and market share. It is well known that success feeds upon itself; similarly, one premium often generates another, with the result that the second million is easier to make than the first.

Some years ago there were what have been referred to as the scotch wars. Competing liquor companies fought to increase their share of the market for Scotch whisky. Dewar's, an old and venerable brand, refused during these wars to lower the price for its scotch. It initially lost some of its share of the market, but did not panic and kept up its image campaigns that positioned Dewar's as the quality scotch, at the high end of the price scale for the product. During this period, one of Dewar's major competitors, Black & White Scotch, became a vigorous discounter. By the time the scotch wars ended, Dewar's had regained its lost market share, and actually "won" in two ways. First, because it had never lowered its price, it continued to make money during the wars, due to its premium. Second, it made even more when the marketplace established, with the help of Dewar's image, the difference between a commodity and their own premium product. And Black & White Scotch had almost completely disappeared.

Consequences of Being (or Becoming) a Commodity

In the future, some products and services that command premiums today may slip and become commodities, despite energetic work with their images. This has already happened today in two important industries.

Credit cards were invented by American Express and

Diner's Club. In the beginning, a customer could obtain one of these cards only if his yearly income was higher than the national average, and his employment history was stable. Obtaining credit was a privilege, and having a credit card became a status symbol. The cards became core businesses for these two concerns.

Then banks began issuing credit cards, principally to induce their own customers to use their credit, which, of course, resulted in increased profits for the banks. In time, bankcards became an enormous success, and the greatest source of profit for such giants as Citibank, the largest single issuer of Visa and MasterCard plastic. Now nearly every family has one of these cards. As important, every bank must offer them as part of their basket of services. Standards for credit eligibility have been lowered, so that a credit card per se is no longer a status symbol. Unchecked, credit could have become a commodity. It is only through image development that it avoided that unprofitable fate. American Express, with its green, gold, and platinum offerings, has managed to maintain a premium image. Visa and MasterCard prefer images of somewhat lower status, but still maintain a premium image based on their service, so that they, too, avoid commodity status. A new entry into the field, the AT&T Universal credit card, has as its core appeal the extraordinary reputation of AT&T.

Another example. There was a time when CBS stood for quality in entertainment, in news, and in other aspects of being a television network. That network had a distinct reputation, and NBC and ABC also had specific images, though the latter were not able to command the same high rates for advertising that CBS charged. Nowadays, the public has decided that the programming of the three networks is pretty much interchangeable, that there is very little difference between a sitcom on NBC and one on CBS, between the evening news as anchored by Peter Jennings or by Dan Rather. The reality of the sameness of their product overwhelmed the networks' images. As a consequence, the particular image of a network no longer matters—nor can it command a premium. The best evidence for this is the fact that in recent years CBS, NBC, and ABC have all shown declining profits and have either been bought up by other companies or have radically changed their managements. The only network to successfully maintain an image different from that of the big three is PBS, the public broadcasting network, which *must* maintain its distinctiveness in order to attract the viewer donations that are its major source of funds.

And what of the future? Just recently, a major gasoline retailing company has determined that in the relatively near future, gas for automobiles will become a commodity. This will mean that the customer will perceive no difference among competing brands. Imagine for a moment how this could affect society as we know it, what gasoline retailing would be like if there were *no* gasoline brands, and no premiums could be charged for specific brand names.

Anybody's Gas

The local station won't have the name Shell or Sunoco on it, but, rather, the name of the area or the retailer—say, Dix Hills or Joe's Gas. For that matter, the station will differ from today's stations in other ways: it will not only be brandless, but serviceless. Since consumers no longer think they're giving up anything when they're unable to buy their favorite brand of gas, there will be no premium that comes to the retailer for service, and, therefore, no reason to offer service. To buy the gas, the consumer will use a regular credit card. Since all that such stations will offer is gasoline, they can be rather small, and the only limit to their number will be the availability of sites, the storage space for the big underground gas holding tanks, and the number of locations to which a refinery can deliver the product. You could conceivably see a gas pump on every corner.

To me, one of the most interesting points about such an analysis is that when a premium product drops to the point of being a commodity, service also disappears. In the present, if you give poor service, and that service is linked to a brand, the consumer has someone to blame and to hold accountable. Bad service at Department Store A means a customer may walk around the corner to Department Store B to buy similar merchandise instead of purchasing it at Store A. Similarly, receiving an invasive "cold call" from Brokerage House C may result in a customer's unwillingness to do any future business with that brokerage, and receptiveness to a less annoying sales pitch from a competing brokerage. So: an intangible value of brand image is that it imposes standards of service on the brand owner, if for no other reason than that the brand tells the customer whom to blame if something goes wrong with the product or service. Brands beget accountability. While brands are frequently vilified by consumers as wastefully expensive, in many respects they are actually the consumer's best friend.

This analysis brings me to an imperative: *To maintain the premium edge when the product is in danger of slipping into commodity-hood, make sure your true value is recognized by your customer base, even if this means improving service and accountability!*

Beating the "No Premium" Blues

Here are some other things that can be done when it appears likely that your premium product or service is about to become a commodity.

1) *Create a new product with an identity separate from that of most other products in the field.* Recently, Lippincott & Margulies created an identity for a new gasoline for Sunoco. Mindful of the notion that gasolines may soon become commodities, and also aware of the public's increasing concern for the environment, Sunoco designed a new gasoline whose emissions will be nonpolluting. It is called Eco-Clear, and the hope is that it will command a good share of the market because it stands out from the pack and has qualities for which the public will agree to pay a premium.

2) *Do some creative repositioning of the product away from its basic reference as a commodity.* Once upon a time, Marlboro cigarettes were dying on the vine because of their image as a feminine brand, until a campaign was undertaken to reposition them as the ultimate masculine brand, via the Marlboro cowboy. Arm & Hammer repositioned baking soda—really, the ultimate commodity—as a refrigerator deodorant, a move that staked out new territory for an old and tired product.

Such a gambit is not always successful. In 1947, more people recognized Borden's symbol Elsie the Cow than recognized Harry Truman, and he was president of the United States at that time. But Borden wanted to position its whole-milk product as superior to all other milks, and mounted a campaign to say so. It failed miserably, because the public perceived that milk was, indeed, a commodity, and rejected the idea of paying a premium for Borden. So disastrous was this campaign that it affected many other aspects of Borden's consumer business, and today a consumer cannot buy Borden's milk or ice cream. Borden also shot itself in the foot (the hoof?) by continuing to provide supermarkets with milk for their private labels while trying to market its own brand. There was no difference between the store brand and the Borden brand, and consumers quickly recognized that fact and bought the cheaper product.

3) *Hang on to a technological innovation.* The can industry was sinking to commodity status a generation ago, when it was temporarily rescued by the invention of the pop-top, the self-opening soft drink and beer can. This novelty exhausted, however, the industry could no longer command a premium for its products, the profit margins went down, and the major players were taken over by outside concerns. I've already told part of the story of the American and National Can companies in the introduction to this book. Because the can companies accepted the notion that a can is a can—that the can was a commodity, not worthy of a premium—they failed to continue to develop meaningful technological innovations that might have preserved the specialness and usefulness of cans (and of can company managements).

One of the pervasive myths of American industry is that if a business requires a great deal of capital in order for a company to become a player, that business is safe from competition. Believing this, and knowing that their business was quite capital-intensive, the can companies thought themselves immune from attack—and steadfastly refused to do anything about their declining images, thinking that images were irrelevant in the fight so long as they were protected by the need for big capital expenditures on the part of any competitor.

Suppose, for a moment, that they had not ignored image as an asset. Before the industry sank into commodity status, what might a can company have done to protect the uniqueness of its image and its product? One possible position: if your soft drink comes in a can, you know your cola company cares about your safety and product freshness. Or: the modern way to drink soda is from a can, not from a bottle. Or: The can companies might have mounted a well-organized and well-funded recycling program aimed at repositioning the can as environmentally preferred packaging. Such an effort might have improved consumers' attitudes toward cans by making them feel good about using them, rather than just viewing them as containers that will make the garbage heap bigger. Such an approach might have staved off the demise of the leading can company managements for some more years.

*   *   *   *   *

So much for the dangers of becoming a commodity. In the next chapter, I'll tell the story of three companies that each, in its own way, came to an understanding of Weill's Law, and for which Lippincott & Margulies helped fashion identities and images to take advantage of their premium status.

CHAPTER 6

Classic Identity Programs

Under the leadership of Walter Margulies and Gordon Lippincott, nearly every year in the 1960s and early 1970s Lippincott & Margulies mounted major programs, each for a very large company in a distinct type of industry, that changed the way Americans viewed these companies and their products. The changes wrought by these programs have become such a part of America's business culture that today it is difficult to imagine these companies and their products without simultaneously thinking of some of their identity components that Lippincott & Margulies helped put in place. Out of the many done by the firm, I've chosen three classics, each representative of the era and a particular industry. Let me briefly outline the circumstances and main points of each, to show how thinking about corporate image can transform a company's conversations with the world.

The Pentastar

When the Chrysler Corporation came to Lippincott & Margulies in the early 1960s, it was a time of crisis and with a mandate for renewal under a new chieftain, Lynn Townsend. Previously, Chrysler had marketed its cars as separate and individual brands—Chrysler, Plymouth, Dodge, and Imperial— each under its own organization. These competed with one another so fiercely that they seemed not to be arms of the same company at all. Going around the countryside, one could find showrooms featuring one or several of these dealerships in combination, but nothing to indicate that all the products being sold were those of the Chrysler Corporation. There was no unifying style in the way the products were identified or sold, though Chrysler products bore an outmoded "forward look" symbol, the so-called double flying wedge. No one knew what it meant, and it looked like something from the 1930s.

Chrysler was the definite third of the Big Three, with only a 10 percent market share, and way behind in more than its corporate style. It also had less money to spend on advertising. For every three dollars spent by GM, and every two dollars spent by Ford, Chrysler had but one dollar at its disposal for promotion of its products. Chrysler needed to do better with less money, and so came to Lippincott & Margulies.

Was it possible, Chrysler asked, to leverage its corporate identity practices in such a way as to make a dramatic impact, and alert the public to its new objectives? Could a corporate identity also function as a marketing umbrella for a number of different brands instead of the single Chrysler brand? Could such a concept be a substitute for spending large amounts of additional advertising dollars?

Lippincott & Margulies made many recommendations to change Chrysler's look and brand identities. We took pains to establish what the field later came to call an "umbrella" identity, one that would subsume all the brands under a single corporate identification, and that would convey the size, strength, and organization of the parent concern, Chrysler. We concluded that words were not enough. Chrysler needed an abstract symbol that would be compatible with any word or set of words, a symbol that would fit in anywhere and would always convey the same message. We created a new symbol for the corporation, the five-pointed solid star that we dubbed the "pentastar." A mark of high-quality engineering and of superior service, this symbol became the carrier for the corporate identity that Chrysler wished to transmit—seen on all signs, in a small but tangible way on the door of every car of every brand, and of course on all communications such as print and television advertising. Subsequently, we designed dealership signage and even entire new showrooms to reflect the notion that Chrysler was the corporate parent of all four brands; even if the dealership sold only Plymouth and Dodge, the Chrysler name and new logo went on its outside sign, with the brands listed below it.

This program had a tremendous cohesiveness. Each car

brand had its name rendered in a particular typeface and in a particular color—the Dodge signs were red, Plymouth's were blue. All the cars bore the pentastar symbol in the same place, on the door. Showrooms were given special looks that linked them to one another. And the Chrysler name and pentastar were splashed across the country on signs, water towers, and billboards, as well as on the trucks that transported new Chrysler cars to the dealerships, and so on. Chrysler the company had been positively separated from Chrysler the brand.

Now, for this program truly to be effective, it had to be embraced—and, in part, paid for—by the dealerships throughout the country. Many dealers had objected to the new idea in outline, but when it was complete and ready to go, management made a tremendous effort to get the dealers behind it. Some had to make extensive alterations to their showrooms. All had to repaint signs, put up new types of poles and beacons, and agree to reprint their stationery, forms, and other communications to conform to the new corporate identification program. As important, many dealers had to acknowledge for the first time that they were in fact Chrysler dealerships, even if they sold only Plymouths or Dodges. Taken as a whole, the program changed the automobile retail scenery, altering forever the way dealers presented themselves to the public.

Soon after the program was introduced, it became the cover story for the April 29, 1967 issue of *BusinessWeek*. The Chrysler identification was splashed all over Detroit so effectively that company executives boasted that Chrysler now seemed to own Motor City. Out in the country, customers and even noncustomers came to recognize the pentastar and to know that it stood for the Chrysler Corporation; furthermore, they accepted the identification system that placed the brands in relation to Chrysler. The image became Chrysler's, not the individual dealers'. Lynn Townsend put it this way: "Customers buy more than a product. [They] buy the company that makes the product. They buy its character, its size, its sincerity, the confidence it inspires. Thus, the function of our corporate identity system is to influence constructively the image of Chrysler as a corporation."

Shortly after the program was introduced, Chrysler Corporation upped its market share to 18 percent. Some months after the debut of the Chrysler program, both General Motors and Ford adopted their own new identification programs, and we have to go along with the old saying that imitation is the most sincere form of flattery. Even Lee Iacocca, then at Ford, thought the Chrysler corporate identity program so good that he later wrote in his autobiography that it was one of the few aspects of Chrysler he admired before he got to the company. Later, after he took over Chrysler, Iacocca took actions that further reinforced this identity program, placing the pentastar as a hood ornament on all of the company's models. Today, the idea of one grand corporate symbol such as the pentastar, and the allied idea of aligning individual brands (each with its own typeface, color, and other identifying marks) under that umbrella identity, are the norm for many great corporations; we're pleased about that, and about the way that Chrysler's innovative system and symbol have withstood the test of time.

Another certification of the identity system's worth came in the mid-1980s, when Chrysler acquired certain high-technology and financial services companies, partly in an effort to improve its stock price through diversification. The goal was to increase non-automotive earnings during the almost inevitable cyclical downturns that affect all carmakers. As it expanded into such areas as aerospace, management needed to reassure employees, dealers, investors, and customers that it would continue to be principally devoted to automobile manufacturing and service. A new corporate structure was needed to reflect the new diversity; it would now be along the model of a parent company and its subsidiaries, rather than a single company structure solely involved in car and truck manufacturing and marketing.

This was a subject of some delicacy. It came at the moment of Lee Iacocca's great popularity as the man who turned the company around, and there was some fear that if it became known that we were working on the project, that would be a signal to the dealers that Iacocca was going to take the company in a new direction and de-emphasize automobiles. So we were kept away from headquarters, had to submit invoices under a code name, etc. The culminating event of the insistence on secrecy came when we presented our recommendations, not at corporate headquarters, but in private to the chairman and his key executives at Iacocca's country club. Some in attendance wanted to adopt a wholly new name for the new parent; they argued that since the multiple for automobile companies was low, and since the company was well diversified, it would benefit from an identity that clearly and unambiguously described a company that was involved in much more than just automobile manufacturing, and that such a new identity might help avoid the curse of the low multiple. But our research revealed that this would impact badly on the company's traditional customers and dealer base. The fact was that regardless of what different companies Chrysler might acquire in the future, the sheer size of its automobile operation would almost certainly dominate its revenue, and a completely new name might weaken rather than strengthen the links that bind such divisions to the parent. Furthermore, the new, non-automotive divisions felt they would be greatly strengthened in their marketing efforts if they were known to be under the Chrysler umbrella, and *wanted* the Chrysler name on their particular "dealerships." The more closely they were identified with folk-hero Iacocca, the better. We recommended keeping the Chrysler name and gradually redefining it to make it stand for more than the manufacturing and distribution of cars and trucks. We developed the parent identity as Chrysler Corporation, and arrayed several "mezzanine operating groups" under it—Chrysler Motors, Chrysler Financial Corporation, Chrysler Technologies, and Chrysler Aerospace, each with separate color coding and nomenclature systems.

Over time, the plan envisioned a communications structure in which all communications would come from the parent, Chrysler Corporation. If the company bought a bank, the press release would begin, "Chrysler Corporation purchased . . ." Or if quarterly figures were to be released, the statement would say, "Chrysler Corporation today announced . . ." However, if a new automobile model were to be introduced, such an announcement would come from Chrysler Motors, or if there was a new financing rate the issuer would be Chrysler Financial. Over time, the differences between parent and subsidiaries would become understood by the public, and so would the fact that the Chrysler Corporation was more than just an automobile manufacturer. Had Chrysler Corporation stayed on track with its diversification, this plan would have been effective, but, after some years, the corporation has abandoned diversification and refocused its efforts on automobile and truck manufacturing. It bought American Motors, and the acquisition of the Jeep brands fit neatly into the existing brand architecture and communications structure that we had already designed. Moreover, the parent-and-subsidiaries structure that we had devised allowed Chrysler Corporation to sell its various non-automotive companies without requiring major repairs to its corporate identity.

In the Blue Box

In the early 1970s, Howard Clark was the chairman of American Express, and he had a problem: the future direction of the company. It was clear to him, since he had planned it carefully, but he wasn't sure if the company's name and identity practices would "work" to help in achieving the great future he envisioned for the company. Would that there were more top management executives with a sure understanding of where their companies were heading in the future—but for the moment, let's concentrate on American Express.

The name had come out of the baggage forwarding service of the railroads. In the United States, the term *express* had superseded its literal meaning, but in Europe—a major focus of the proposed expansion of the company—the term still connoted such diverse things as a train, a newspaper, and a laundry service. Moreover, the idea of something being called "American" was not necessarily good for business in Europe or in Asia, where the image of "the ugly American" was still in vogue.

Clark was more concerned about the name being inadequate to fit the future strategy of the company. At that time, American Express included an international travel booking service, financial service products such as the American Express credit cards and Travelers Cheques, an international bank, and Firemen's Fund, an insurance company. In the future, Clark wanted American Express to become a financial services giant, a supermarket able to provide many synergistic money-related services to its affluent customers, and to do so on a truly global basis.

We sent people throughout the United States, Europe, and Asia to interview travel agents, retailers, hotel managers, customers, and potential customers who might need the services of American Express. In Europe, for instance, we found that the company was universally looked upon as the American travel nanny, the company that took care of Americans when they went overseas, providing mail and money drops, and so on. This was less than Clark wanted. Another problem was that acceptance of the card at retail locations was not particularly good: some merchants would beg to give discounts for cash rather than have to bother with the paperwork of the credit card transaction and pay a significant commission to American Express.

On the other hand, the name and the American Express card had come to symbolize membership in an affluent, international elite—and that was something on which to build.

While we were interviewing, of course, we found and photographed the signs and establishments in which the company did business, and discovered that they were incredibly confusing. The visual expressions of identity were often outdated and frequently fragmented. In some places the name was shortened, translated, or otherwise distorted, and there was no single system of identification that linked the disparate entities.

We learned that the travel agency business itself was a money-losing proposition. If so, we asked, why did the company stay in that business? Because, Howard Clark answered, the travel agency business was the glue that held it all together. The travel agency provided assistance on where to go and what to see, provided the credit cards and checks, steered people to the banks, and so on.

The company presumed that we would recommend a name change to align its future—such a recommendation would obviously be in our own self-interest—but we explained that our research had revealed that the old name was one of the company's strongest assets. Lippincott & Margulies believed that the name American Express could be extended to represent properly the diverse elements within the company, and that when redefined it had the potential to become a fine name for the future financial services giant envisioned by Howard Clark. To organize and regularize the company's communications, we designed an entire nomenclature, logo, and signage system. There would be a little blue box within which the American Express name would be positioned and silhouetted in white; everywhere, this symbol would convey the identity. Travel bureaus, banks, and other American Express locations would bear the distinctive box and newly designed typography. Placed in all locations, and instantly recognizable, the blue box—some referred to it as a "blue chip"—was a sign of service that shortly became ubiquitous. A simple but telegraphic symbol, it was used to endorse and promote a wide array of products and services, including travel, credit cards, investments, international banking, trade, finance, and financial planning services. It became one of the most widely recognized corporate symbols in the world.

The universality and effectiveness of the symbol and the entire nomenclature and typographical system are well established today, but in the early days when Lippincott & Margulies had recently made the recommendations, they were not so obvious to everyone. In particular, we dueled with American Express's main advertising agency, Ogilvy & Mather. Ogilvy did not want to use the blue box in conjunction with the American Express card in television commercials for the card. The advertising agency argued that the distinctive centurion on the card, and its money-green color, were enough identification, and that adding the blue box would be too much for a viewer to remember. Our position was that if the company was trying to establish an identity that would carry throughout its line of services, it was essential to implant that identity on such widely visible products as the credit card, even if it merely appeared on the back of the card or in a secondary position. Only the intercession of Howard Clark was able to solve the issue: he told the agency to make sure the blue-box logo appeared on the credit card and on-screen in the television advertisements for it, and Ogilvy & Mather then rather quickly acknowledged the wisdom of his words.

Equity Elements

In 1965 the Coca-Cola Company asked Lippincott & Margulies to evaluate its brand identity practices to ready the company for the future. On the horizon was more emphasis on global marketing; in the short term, there was the press of competition from other beverages. They were losing market share to Pepsi Cola and other soft drink beverages. Our job was to determine what role, if any, the visual presentation of the brand might have played in the decline of the brand. We needed to know what sort of an image was being shaped by the way the brand appeared in all its various expressions.

As we did our research, we discovered that in its communications the brand had many looks, not one. The packaging was so different from market to market that the company had difficulty ending its television commercials with the picture of a package, because it didn't know whether the package that appeared on the screen would be available at a particular local point of sale. Although the Coca-Cola trademark and the product it represented had received the greatest exposure of any product in history, the cornucopia of diverse and fragmented ways in which the brand appeared weakened its image. Coca-Cola's signs had been around for so long, and in so many permutations, that they were having what Walter Margulies called the "wallpaper effect," in which something that is supposed to stand out has actually faded into the background. Coca-Cola visuals came in several typefaces, several colors—green and yellow, as well as red—and with dozens of background designs. Some outlets were using signs from the 1920s or 1940s, others from more recent eras. Trucks were emblazoned in a variety of colors, as were billboards. Each gondola or fountain dispenser showed the product in a different way. The reason for the disarray was that as the company developed, it adopted the policy of awarding local bottlers the right to bottle and sell Coca-Cola without charge, but required those bottlers to buy their own packaging and local signage. While the company provided standards for graphics, they were usually ignored by local bottlers, who considered graphics a low-priority item. The prevailing attitude was that the presentation of the brand identity was a local issue, since the local bottler paid for it. The idea of a national brand, let alone a global one, simply had not occurred to local bottlers; so long as the company's advertising campaigns were visible and well funded in their local market, in the view of the local bottlers that was all that mattered.

We convinced Coca-Cola management that in order to achieve its global ambitions, in order to leverage the vast heritage its brand possessed, and in order for the brand to appear as large as it actually was, what was necessary was a disciplined approach to the visual expression of the brand across all media. Packaging, straws, uniforms, cups, vending machines, etc., were important media with which to shape and reinforce the brand's image, and their current fragmentation was actually diluting the appeal and stature of the brand. To limit headquarters' involvement just to advertising was like running a car on two wheels— it would not take them where they wanted to go. We convinced Coca-Cola that the only way to fix its problem was to investigate thoroughly the design practices and create a new, organized design system that would encompass every conceivable way of communicating the Coca-Cola message. That involved going to the heart of the business, then in existence for over eighty years.

Even the name itself had some problems. Back in the 1920s, a small competitor—there had been hundreds, over the years—had recognized that the public said "Coke" when referring to Coca-Cola, and incorporated as "Coke Company of America." Coca-Cola sued, and, while it was suing, mounted a major advertising campaign on the theme, "Don't call our product Coke," a campaign that insisted customers ask for Coca-Cola by its full name; otherwise, all you'd get was an ordinary "cola." The case went all the way to the Supreme Court before it was decided in favor of Coca-Cola; as a result of the suit, though, the company had to be very careful in the way it labeled and promoted its products, and had to diligently enforce its legal requirements on every level possible. Part of the suit decided that the word *cola* had become a descriptive term, and could be used by many products, though Coca-Cola itself was recognized as a registered trademark. Therefore the company had to write special guidelines for when the word *Coke* could be used, and when the full name of Coca-Cola was required; for instance, the

Image by Design

guidelines said that both appellations must appear on a single package.

What Lippincott & Margulies wanted to do was to visually present Coca-Cola right, and to do so in one single, powerful, and memorable way; seen once, it would, we hoped, be remembered forever. Despite years of attempts from Atlanta, local bottlers and regional retailers were pretty much presenting Coca-Cola in whatever way they saw fit. Their vision was limited to the territory they served. The company, recognizing the value of a single brand image, was ready to introduce a single design system and to compel its use throughout the entire organization and by its bottlers, but the question still remained: how should this design system look, and what image should it project? The immediate answer was that, regardless of any weakening of the company's leadership position, in the 1960s there was only one Coke—Coke was it—and the past and the company's history had to be a large part of the formulation of its future.

From the historical plethora of Coca-Cola typefaces, slogans, backgrounds, names, shapes, and other components, our research isolated four "equity elements." These were factors in which the Coca-Cola Company had built up equity through years of use. They were: 1) the flowing script used to depict the company name; 2) the color red, the most predominant background color at the point of sale; 3) the word *Coke*; and 4) the shape of the Coca-Cola bottle. This last element had even been trademarked, a most unusual thing. In the company's prior presentations, these elements had been present, but not coordinated. The consumer had never been exposed to all four in a way that interrelated them.

From these four elements, we created a design system for Coca-Cola that included the flowing script version of the name, the background red color, the word *Coke* (rendered in a more modern script), and, under the name, a swirl that was originally described as a dynamic contour and that came to be called the "wave." This wave is a derivative of the bottle shape; in fact, it's the shape of one side of the bottle. There was some consideration given, earlier in the design and selection process, to having a similar line on top as well as on the bottom, a plan that would have contained the name within the bottle shape, but this was dropped in favor of a simple, single line. So distinctive was this shape that regardless of the language in which the words *Coca-Cola* or *Coke* appeared, the dynamic contour, a sweeping white swirl, would still enable instant recognition of the product.

Great secrecy attended this project, as well. Paul Austin was the CEO, and the legendary Robert Woodruff was chairman of the executive committee and the major stockholder; both saw fit to keep the Lippincott & Margulies program—code name, Project Arden—from the forest of people in Atlanta headquarters for quite some time. In New York, Walter Margulies misled interested competitors and journalists by saying that the firm had been hired by Coca-Cola only for a new product package design assignment.

In the dead of winter, Lippincott & Margulies employees came down to Atlanta to an isolated, guarded warehouse and installed versions of three competing design systems on every conceivable type of display: supermarket aisles and shelves stacked with the product in bottles and cans, large and small delivery trucks emblazoned with the wave and red color and other elements, uniforms for deliverymen, cartons, fountain set-ups, billboards, and packaging of every type, all in great profusion, and all featuring the candidate design systems. A capping machine was brought to the site so that Lippincott & Margulies staffers could actually bottle some of the product—for even the bottle caps had been redesigned with the alternate logos.

Woodruff and Austin walked through the cavernous warehouse alone to size up the effect. Austin, of course, had already seen and approved our recommendations, but Woodruff owned 30 percent of the company's stock, and nothing of this scope could possibly happen without his explicit approval. Woodruff thought it was just fine, but suggested "they" make no changes to the old typeface—so skillful was the design system that he hadn't realized that the entire look of the brand's identity had been changed.

Once approved, the system was sold to the bottlers, and the results were startling. The red Coca-Cola trucks, the signs featuring the "wave," the distinctive typefaces, and the repeated impact of having the elements stand out on everything from the employees' caps to the point-of-purchase displays, from billboards to bottle caps, boosted Coca-Cola's recognition in every segment of the public. The brand was recognized on sight, and didn't even require consumers to be able to read the brand name. Even the alphabet in which the brand's name was recorded became of secondary importance. The look was ubiquitous and organized; each exposure of it reinforced all the others, and the design system became a crucial element in resurrecting the Coca-Cola brand's leadership of the soft drink industry.

In the past quarter-century since the introduction of this design system, Coca-Cola's image and the way it communicates its products have stood both the test of time and the test of global marketing. Recognized everywhere, they have helped sales abroad to become a major factor in the company's continued profitability.

CHAPTER 7

Image: To Buy or to Rent?

Convinced by the success of such classic programs, the compleat businessman or businesswoman thumps the managerial desk with glee, and shouts, "An image for me!" He or she does some research and finds out that images are expensive things to cultivate; moreover, it takes quite a bit of time to build one, and so sometimes management decides that, rather than pay to build an image, it would prefer to go right down to the image store and rent one.

Glamour for a Price

Instant history and instant awareness are the rewards of renting glamour. Typical rented images are those of movie stars or sports celebrities. A recent campaign for an Elizabeth Taylor perfume is an example that comes immediately to mind. An expensive rental! The legendary Liz demands and receives her pound of flesh, so to speak. Not all such rented images come attached to a person or need be paid for at quite so steep a price, even though many can yield equivalent dividends. Some are even free: an alert General Motors executive named a car Monte Carlo, and the company reaped rewards in the form of the glamour associated with the name of that city without paying thousands of dollars for the privilege of using it. Pontiac was once the name of a tribe and a city; today, of course, it has become so synonymous with the line of automobiles that bears its name that few remember its earlier association.

Today, though, most rented images entering the marketplace have to do with celebrities. The less expensive ones have long since been snatched up. Since Elizabeth Taylor is an extremely well-known and alluring woman whose public image matches the imagined associations of a fragrance, a perfume line bearing her name, Elizabeth Taylor's Passion, provides many advantages to the renter, not the least of which is the "free" publicity attendant upon everything she does or says. In this instance, the deal between company and star also included such valuable assets as Ms. Taylor's participation in the advertising and merchandising campaigns. In theory, the name alone adds value, and the availability of the living, breathing persona attached to that name should represent even greater added value.

Perfume makers often resort to rented images, principally because building awareness and recognition takes so long and costs so much. On the other side of the ledger, fragrance merchandisers have learned that the chances of success are greater when a known market exists for a particular name. Calvin Klein's Obsession was recognized the day it was introduced; it took Giorgio three years to launch on a national scale. The theory suggested that the cost of launching Elizabeth Taylor perfume would be less per unit than these, because the recognition level of Ms. Taylor was already in the stratosphere. The son of the owner of Fabergé, who also wanted to market a new perfume, believed that it would not sell well if it bore his own name, and so paid Mikhail Baryshnikov and named it Misha. Other well-meaning attempts also have failed: Cher and Catherine Deneuve, to name just two. Clearly, awareness and popularity are not enough to sell a product.

The recognition-by-name process is not limited to perfume. Gloria Vanderbilt's name adorned jeans made by Murjani, whose own foreign-sounding name would not have brought with it the cachet associated in the American public's mind with the socialite heiress.

In 1990, the jeans manufacturer with the unlikely corporate moniker No Excuses hired Marla Maples, the gorgeous self-admitted "other woman" in Donald Trump's life, as spokesperson. For lending her name and curves to the product, Maples reportedly received $600,000; but regardless of the appropriateness of her reputation to the product, the whole idea of latching on to a tabloid queen whose media visibility has to be quite short-lived represents a low in short-term thinking. No

Excuses will have to sell a lot of jeans very quickly to recoup its investment. As Maples' star dims, as it surely will, her value to the company will just as surely fade, and I don't think built-in fading was part of the makeup of these particular jeans' fabric.

This points up the prime disadvantage of rented images: movie stars and sports celebrities, being human, are fallible—and some are quite a bit more vulnerable than many lesser-known human beings. They are prone to unpredictable behavior that often brings with it embarrassing results. If your rented movie queen puts on weight and enters the Betty Ford clinic, do you want your product to evoke an image of fatness and alcoholism? Of course not, but that is a risk that comes with associating your product with a fallible human. Do you now pay for your rented star's rehabilitation in order to maintain the image on which your product is riding? Of course not; you dump the star, because it costs too much to change the rented image once it starts to sour. However, had the image been entirely your company's own, and there had been a problem, you would have had a greater incentive to work and spend to resuscitate it. Leona Helmsley seemed a perfect image and spokesperson for the Helmsley and Harley hotel chains controlled by her husband—until the day she stood accused of nefarious deeds. Then her public persona was considerably tarnished, but the amazing thing is that for some reason—ego, assuredly—the print advertising campaigns for the Helmsley and Harley hotels continued to bear her name and photo, and to make reference to "the queen" standing guard. Now that she has been convicted and sentenced, will the company hotels be able to recover their own image and dump hers, seeing that the Leona Helmsley name has become associated with criminal deeds? Six months after her conviction, Mrs. Helmsley decided that, in the words of her spokesman, "The hotels are bigger than any one person" (*New York Times,* February 28, 1990), and that she should no longer appear in advertisements for them. It remains to be seen whether this tardy decoupling of the queen and her domain will help the hotel chain.

At the outset of his career, Donald Trump made a decision: to create a glamourous image for himself that would become the heart and soul of his commercial enterprise. This image was based both on the perception of larger-than-life wealth, and of personal attractiveness. To quote one of his spokespersons, "When you buy in a Trump building, you're buying sex and dreams and excitement and a lifestyle" (Blanche Sprague, quoted in *Newsweek,* March 5, 1990). This represents a deliberate attempt to equate Trump the person with his products. But problems arose when one of the pillars of Trump's image was brought into question: his absolute wealth is not nearly as absolute as it was believed to have been, and this, in turn, has caused downgrading of his personal image. And, as I have noted just above in connection with Marla Maples, the other pillar, Trump's personal style, simultaneously came into question. When the details of a messy divorce were raised in newspapers and magazines, the linkage that once seemed helpful became compromised. Through the controversy, the awareness level of Trump the human being soared to unprecedented heights, it is true, but damage was done here, too, especially when Trump rhetorically asked, in a television interview, why, in the prime of his life, he ought to be married. It was fun to fantasize along with Trump, but then the business public began to look askance at someone who disregards a cherished American value, and who seemed as if he was going to be chintzy in a divorce settlement with the mother of his children. A downward influence on the image of Trump's products and services was inevitable. In saying this, I pass no judgment on Trump's personal life-style; my comments only highlight the risks attendant on relying for an image's sake on human beings who are alive—and unpredictable.

Remember Roy Rogers? The once-popular singing cowboy lent his name, his image, and for a time his personal involvement in advertising campaigns when the Marriott Corporation opened its first fried-chicken-and-roast-beef-sandwiches restaurant in Virginia in 1968. The chain grew in size until 363 Roy Rogers restaurants existed under the Marriott umbrella and 238 more were franchised. By the late 1980s, however, even the cowboy's son had to admit publicly that most of the young people who ate fast food in the RR restaurants didn't know who Roy Rogers was. In early 1990, Hardee's completed a deal with Marriott to buy its 363 Roy Rogers restaurants for $365 million—and announced immediate plans to drop the Roy Rogers identification and make them into Hardee's, which sold the same kind of fast food anyway. The price for the chain would have been higher, many people said, if the cowboy's name and persona still retained the sort of star quality that had attracted customers to "his" restaurants thirty years earlier. Roy had a long ride in terms of fame, but his image no longer "worked" once those who were young when Westerns were wonderful outgrew fast food and passed on to more sophisticated grazing.

Who's Speaking for You?

It can be argued that, like Leona Helmsley and Colonel Sanders, personal corporate symbols *always* get into trouble when they confuse their position as spokespeople with the notion that they are the company's flag. Chrysler's Lee Iacocca understood this trap, and in the commercials in which he has appeared, he has steadfastly viewed himself not as the embodiment of Chrysler, but rather solely as its spokesman. His dialogue always is about the company rather than about himself, and focuses in the strongest possible terms on the overall Chrysler image. Iacocca's credibility does not come from being a celebrity—from how he dresses or looks—but, rather, from what he does for a living, from his professional craft, which is to make and sell cars. When Iacocca went through a divorce, because he had separated his own personal image from that of the Chrysler company, there was no damage to Chrysler. The only minor misstep was an attempt to extend his own image to a brand of olive oil; that failed. This mishap aside, when the time comes for Iacocca to retire, because of the work done to separate the man from the car company, the Chrysler image will endure for the company as a separate asset, one that is not forever or inextricably linked to Iacocca's person.

And then there was Malcolm Forbes. He was not only a spokesman and ambassador for his magazine, but the principal owner and a man devoted to its success. However, Forbes the man carefully and shrewdly institutionalized the magazine that bore his name, and nurtured it until it had its own image that was quite separate from his own. He, personally, was associated with motorcycles, hot-air balloons, and stylish parties in exotic places; his magazine stood for insightful business journalism and lists of the most successful people and companies. Forbes even institutionalized his selling tools; his famous dining room on Twelfth Street in Manhattan remains *the* place to have lunch, even after his death, and his yacht is still a glamorous place to be. As a consequence of this separation, there is life for *Forbes* magazine after the untimely death of Malcolm Forbes.

Beyond the idea that human beings are fallible is the truism that he or she who is currently hot may soon be not. What goes up quickly quite often comes down just as quickly, and if you have harnessed your product in tandem with a particular rock-and-roll singer, for instance, you may regret that decision when the person's band disintegrates or fails to follow up the first big hit with another, and falls off the charts into oblivion. If you feel you absolutely must have Michelle Pfeiffer or Tom Cruise—okay. But use them only in an advertising campaign, and never permanently fuse their identity with your own. The public will buy Paul Newman's Own brand popcorn, a specialty product associated with the movies as he himself is (and whose corporate profits are donated to charity), but would not be inclined to snap up Paul Newman's Own motor oil, marketed by Exxon.

Can I Just Borrow It for a While?

To avoid some of the built-in problems of harnessing your image too closely to a fallible human being, you can link your brands' image temporarily to a celebrity through advertising. Rather than naming a perfume after a movie star, hire Catherine Deneuve to speak for it; hitch Coca-Cola to "Mean" Joe Greene for one of the most memorable commercials of the television age; let suave Cliff Robertson be the one to convince consumers to stay with AT&T long-distance service. Pepsi Cola recognizes that linking its image with a star is essential, and further realizes that stardom is ephemeral and that each new generation of potential heavy Pepsi drinkers must be courted by a star who appeals to them—and so hires Michael Jackson for an expensive campaign, using the language and persona of this particular teenage idol as a way of making an old (and unchanged) product seem new and exciting and part of the youngsters' culture. The bet is that once these youngsters have been hooked on the Pepsi habit at just the right time in their lives, they will find it difficult to kick.

In my view, the practice of linking celebrities with a company's image is overdone, and the easy way out—the way, incidentally, too frequently favored by most advertising agencies. Did Hertz really need O.J. Simpson to sell its products? Does the Beef Council gain credence it could not otherwise obtain from commercials featuring Cybill Shepherd? In another era, did General Electric require the services of Ronald Reagan to convince us that progress was their most important product?

Frankly, there's rarely a need for image borrowing. In the main, it is motivated by a lack of creative ideas on the part of the ad agencies. By putting in a star, an ad agency appears to be doing its job, when it may actually be avoiding the real task. Commercials with celebrities may be expensive and easy to sell to a client, but since in them the celebrities are seen out of context, they do not build the sponsor's image in any permanent sense. Nor do they build it—and this is the crux of the matter—in any *proprietary* sense, in a way that will remain with the company long after the temporary association with the celebrity has ceased. In such commercials, the popularity of the celebrity is seldom transferred in any good or lasting way to the product or the company, and the only beneficiary of the process is the celebrity, who is probably laughing all the way to the bank. Well, maybe the ad agencies are beneficiaries as well. For the agency, using celebrities in ads carries with it the possibility of sidestepping blame for wrong headed campaigns: "How could I know that this star wouldn't sell soap? Look at the Q-ratings that show how popular she is."

Confirmation of my view comes from the *Wall Street Journal* of July 5, 1989, which reported a study by Video Storyboards. Sixty-four percent of the viewers polled are reported as saying that the stars who have just performed in the commercials are just doing it for the money—as opposed to the only 16 percent who believe the celebrities are truly involved with the products or companies they are speaking for and who perceive the celebrities as persuasive.

If you run risks by permanently or temporarily linking your product or company to a celebrity, what other solution is there? To get at the answer, let me ask the same question about another corporate asset: if you rented office space, would you spend money on capital improvements to it, improvements that you couldn't take with you if you moved? Of course not. You'd wait to make improvements to office space that you either buy or own or will be using for the next ten to twenty years through a long-term lease. The image asset should be considered in exactly the same way. It is important that the company's resources be put only into those image-related activities that buttress the proprietary aspects of the image, not the short-term rented ones.

Create an Image from Whole Cloth
I am sometimes asked what is the most unique corporate product image, and my answer is Mickey Mouse, whom I call the ultimate image brand. Notice I didn't say "brand image." When Walt Disney needed more than himself to accomplish what he wanted to do—build an animation empire in Hollywood— he created an image from whole cloth, one that had no basis in reality. Mickey Mouse, because he is a created character, has none of the problems often associated with a rented image. He can never be a source of embarrassment. He can't show up in inappropriate places unless the company is dumb enough to allow him to do so. Notice that Disney Studios will allow Mickey Mouse to adorn a watch, but not a cigar or a bottle of vodka.

Some exposures, though they might bring in immediate cash, would not burnish the image in terms of its longevity and utility in other endeavors, and so are rejected outright.

So long as Mickey is not put to churlish uses—cigars, vodka, machine guns—he cannot be overexposed, which is another danger that rented images often court. (However, a licensed dessert product marketed by Nestle in Europe, called Mickey Mousse, is precariously near the brink. One shouldn't permit cute differences in spelling; they threaten the integrity of the trademark.) As for overexposure, there exists the possibility that Charles Schulz's Snoopy, a terrific invented character, now part of the advertising for Metropolitan Life as well as the star of his own line of greeting cards and many other products, is flirting with danger in this regard. But Mickey Mouse, wonderfully charming, and seen as an embodiment of childhood at its best, kindest, and purest, is a great character to have representing the Disney company. As an aside, let us note that part of Walt Disney's genius, beyond creating the character, was in allying his creativity with the sort of people who could merchandise the character in the best way, and in closely and continuously controlling how this image was to be managed. Disney fiercely protects every use of its created characters. When at the 1989 Academy Awards a woman dressed as Disney's Snow White cavorted a bit seductively in an opening dance number, Disney executives were aghast at this threat to the virginal image of their copyrighted character. The studio threatened a suit until the Academy of Motion Picture Arts and Sciences apologized for the unauthorized (and unseemly) use of Snow White. A passing note: Disney did not own the original Snow White, a character in a fairy tale, but through the brilliance of the execution of the animation, and the copyrighting and patenting of Disney's version of that tale and those characters, arrived at a position of actual ownership of an image that is crystal clear in the public's mind.

The proposed purchase by Disney Studios of the late Jim Henson's Muppets would have been, in terms of image, a brilliant match, since it would have brought to Disney more "image brands" to maximize in Disney's well-established ways, and would have given to the Henson company the strength in merchandising that it might otherwise not have been able to develop. But the nuptials were called off.

Another invented image, also enormously successful, is Ralph Lauren's Polo symbol. Today, Polo has been extended to a whole range of clothing products for men, women, and children, and even to bedding, furniture, and other objects for the "country home." The way in which the public has taken to Polo clothing and everything else bearing Lauren's image is testament to the power of the icon he has chosen; Polo has made him one of the country's wealthiest men.

Polo is everything Lacoste's crocodile is not. More than a sport, polo is a way of life reserved for aristocrats and aristocrats only, now and in the future. It takes a lot of money to play, and always will. Polo connotes more than the shirt that the player is wearing, and its image can encompass the entire upper-class, leisure-oriented scene. Polo isn't baseball or bowling, and it isn't tennis, which was once an upper-class pastime but has in the modern era become a sport of millions. Polo, though a vigorous endeavor for its practitioners, won't become a sport of the masses, and, therefore, the image of Polo will remain undiluted.

Items that bear the Polo label are known to 1) always be expensive; 2) be of top quality; and 3) fall within a narrow range of styles. This latter aspect is important, for while there are other American designers of equal or possibly greater stature—Bill Blass and Calvin Klein come to mind—their styles and tastes connote broader ranges. One cannot always predict (or be assured of) what a Blass or Klein creation will look like, but the potential buyer of a Polo-labeled product always has a firm expectation of its probable appearance and what impression it will make upon other people. One further dividend of this image: due to Lauren's skill and credibility as a tastemaker, he can also establish new trends within his admittedly narrow range of styles. A customer walks into a shop featuring Polo merchandise and murmurs to himself, "Hmm, I didn't know you could wear that sort of tie with that color/pattern shirt—but if such an arbiter of style as Lauren says it can be done, I must accept the possibility, and perhaps try it myself." These possibilities have as their basis the brilliance, specificity, and wide range of the Polo idea. To have products that sell very well, are identifiable yet upscale, and a line that is continually broadening to include more and more—what more could one ask from an image?

Don't rent—invent.

CHAPTER 8

Present at the Creation

One of the most exciting times to be in business is at the start of something big. Lippincott & Margulies has twice had the pleasure of being present right from the start of creation. The first was a troubled but momentous birth.

Breaking Up Is Tricky to Do

In 1983, as a result of a court case that took years and many millions to settle, Judge Harold Greene (in his infinite wisdom) decided that the total monopoly of AT&T needed to be broken up. At that time, AT&T was considered the world's finest telephone company, providing the United States with a system that was the envy of all others and the model for most of them. AT&T certainly was a monopoly, since it was a manufacturer of telephone equipment and the supplier of all telephone services such as local and long-distance calling, but it seemed rather benevolent to most people. Not for nothing was the company popularly called Ma Bell. There is an old adage: "If it ain't broke, don't fix it." The judge appeared not to have been apprised of this sound advice, and so he proceeded to dismember the monopoly.

He mandated seven regional companies, separate from AT&T, which was going to be allowed to maintain its status as a manufacturer of telephone equipment and a provider of long-distance telephone services, but would now be subject to competition from other equipment manufacturers, and, for the first time, from long-distance service providers like the fledgling Sprint and MCI services. AT&T had anticipated it might be broken up, and petitioned the judge to retain sole use of the bell symbol and Bell name, and to call itself American Bell. In a part of his decision that has interesting implications for the case I've been making for the value of brands and trademarks, the judge quashed AT&T's attempt. He said that the bell and the Bell name were valuable trademarks, and that they could be used by the new regional companies; if AT&T used them, the effect would be to vitiate the breakup, since there would still be a Ma Bell, only under a slightly different name. Therefore, AT&T would have to give up the logo and the Bell name, and continue on in business as AT&T.

As I've related earlier, Walter Margulies was apprised of the impending split up of AT&T by a board member of New York Telephone, and was asked if Lippincott & Margulies would like to design the identity of the largest of the regional companies, whose territory would be New York and New England. I had recently bought Lippincott & Margulies, and this was an opportunity that was welcome and timely, so of course we said yes, and Bud Staley, the designated future chairman of the regional company, agreed to hire us.

Of immediate consideration was the urgency with which a new name was needed. The judge, rising above such practical concerns as time and schedules, insisted that the new name had to be available within twelve weeks, so that on the day of the announced restructuring, shareholders of the old AT&T could turn those shares in for new AT&T shares plus a predetermined amount of shares in the seven new "Baby Bells." Obviously, shares could not be issued for a company that had no name. In order to meet the deadline, my associates and I simply suspended family life and other social activities. Adding to the difficulty of the task was that six other companies were simultaneously going through the same name-choosing motions, and we had no idea what names they were likely to adopt. Names are often positioned against one another by competing or overlapping companies, so this was an important consideration, as was the need to have each of the seven companies' names separate enough so that they could all be legally cleared in time. No one had ever told me that running Lippincott & Margulies would be easy, but I didn't anticipate this high a degree of pressure so soon after I had assumed the chairman's chair. Baptism under fire, anyone?

The seven chairmen-to-be of these regional companies proved to be quite competitive with one another, even though each had a monopoly on telephone service in his own geographic area, because the new companies were going to be allowed to go into unregulated businesses too, and these were to have no geographic limits. I mention this because it is a key to understanding our approach to the identity of the new company, which would be in two businesses: one regulated by the government, local and national public service commissions, and the like; and the other, unregulated. For many years, regulated regional companies such as gas and electric utilities had offered safety to investors and a nice but modest dividend return on investment. Such stocks were always labeled widows-and-orphans stocks, but were not perceived as having much of a chance for capital appreciation. The northeast regional company chairman realized that his new company was going to need an enormous amount of capital. It was going to start life—start!—with annual revenues of $16 billion and 100,000 employees, meaning that it would be at its inception one of the largest companies in the United States. In order to attract that amount of capital, its stock must offer investors exciting possibilities for capital appreciation, and a widow-and-orphan, regulated company image wouldn't do. It was essential for the company to be perceived as more than just a regional telephone provider engaged in a regulated business. The identity, in other words, would have to tell the world that it was also involved in unregulated businesses and would be concentrating on those to try and make big profits. So our first audience for the company's identity would be the financial community and those who would become the company's stockholders.

There were two additional critical audiences to be considered as we fashioned this new company's identity. Number Two consisted of those 100,000 employees who were disoriented and confused by the judge's breakup decision. For years they had worked for Ma Bell, and had enjoyed the security that that nickname implied; now they were all of a sudden going to be working for an entity that no one had previously heard of. They didn't know what they were in for, and needed to feel secure in their jobs and in their relationship to the new company. Number Three were the local telephone service consumers, who were even more confused than the employees. To whom would they write a check? To whom would they complain if service were bad or interrupted? Most Americans didn't understand why AT&T was being broken up, much less what it would mean to them in terms of their own telephone service.

With these audiences and criteria in mind, we recommended that the new company be called NYNEX: the N and Y for New York, the N and E for New England, and an X to indicate and connote high technology. Our new word, though not in the dictionary, was easily pronounced, and had other important attributes. For instance, it was easily justifiable to those 100,000 employees as reflective of the regional character of the company.

Some people asked why we did not recommend some name that included "Bell," since we thought it advantageous for the regional companies to have use of that name and symbol. In fact, we heard rumors that some of the other regional companies were going to have names that included the word *Bell* and that were also quite regionally specific. Going back to our consideration that the new company needed to be perceived as being in unregulated businesses, we rejected both the bell—symbol for regulated services—and the overtly regional sort of names. This decision later proved prudent when Southwestern Bell tried to market telephone books in the NYNEX area; it was just not credible for a company perceived as being from cowboy territory to sell yellow pages in the canyons of steel of the Northeast, and so Southwestern Bell's attempt failed, and clearly its misnomer (in the eyes of Northeasterners) hastened the demise of this particular product.

The new company needed identity practices to differentiate its regulated businesses from its unregulated ones, and to establish separate yet linked identities for an entire range of the unregulated ones. We recommended retaining the names of NYNEX's regulated subsidiaries New York Telephone and New England Telephone, and making sure that subscribers directed all their attention, payments, and service needs to those entities. This strategy allowed local customers to continue making out checks much as they previously had done, and reduced the feeling of abrupt change in their service or the name of its provider. The bell symbol was used as a prominent part of the logo for both of these local companies.

On the other side, we recommended that the whole line of NYNEX unregulated businesses be positioned so that they all primarily made use of the NYNEX name—NYNEX Enterprises, NYNEX Information Resources, NYNEX Business Information Systems, and NYNEX Mobile Communications. Future advertising campaigns to sell these services would also reinforce the idea that NYNEX was a company that engaged in unregulated businesses and whose stock could offer capital appreciation. Logos for these businesses at first featured a small bell near the NYNEX name, as a bridge to the company's former connection to Ma Bell, and plans were made to remove this last link after the transition period was over.

To start a new company from scratch was a complicated venture, and communications were of the essence. For example, it had been born without even a letterhead. Lippincott & Margulies produced a series of a dozen manuals on how to use the new NYNEX identity on everything from vehicles to signage to advertising. In a final statement of the importance of corporate identity to the new company, we positioned NYNEX's corporate identity as encompassing all of the company's communications—and so some of these manuals that we wrote were actually codes of conduct for the employees, teaching them the manner in which NYNEX wanted them to do business.

The NYNEX story tells of our attempt to midwife into a new existence an already large and distinct company. In the next matter, we were starting, as it were, from conception.

When the Image Was Everything

Nissan USA came to Lippincott & Margulies in the late 1980s and said that Nissan wanted to get into the luxury car market. It had recently conceived—that is, started the model development process, which in Japan takes about three years, and knew it could introduce a new luxury car in two years. Well in advance of the car's completion and manufacture, the company was asking identity consultants how it should accomplish its goal. Right away I knew this was a tremendous opportunity, because most companies that want a new name or logo for a product come to us when their product has already been designed and is ready to be rolled out; that's what had happened, for example, with Cadillac's Allante, and although we were proud of our work on that, my associates and I wished we'd been in on the project at an earlier stage. Now Nissan seemed ready to bring in an identity consultant to be present at the creation. We suggested that the company's new product must be introduced in such a manner that the elements of the nomenclature and design system, and their objectives, were all integrated and reinforced one another. We would give them, we said, more than a name and a logo; we would first help them position the new product and then design an entire image-shaping system as well as ways to manage that image now and in the future.

Nissan liked our all-encompassing approach, and we started to work, together with marketing consultants Canter, Achenbaum. At this point we had no actual product to work with, just a clay model of an automobile. For quite some time, then, the entire product would consist of—an image.

Now prior to this time, the primary target market, American consumers with $40,000 to spend on a new car, had seen Japanese cars as inexpensive products. In the previous decade, Japanese cars had also come to be valued by Americans as of good quality. But there had never been a Japanese luxury car to rival Americans' standard references for such vehicles, the Mercedes, the BMW, the Jaguar. Interestingly, since Americans were already predisposed to accepting a foreign-produced car as a luxury item, this could be an aid to Nissan entering the market. In studying the target marketplace and the luxury car competition, we decided that this new product ought not to be seen as a cheap version of a Mercedes; it would have to set its own standard.

From our interviewing, research, and discussions, we believed that the new product ought to be positioned as something uniquely Japanese, reflective of its several-thousand-year-old culture, its perceived closeness to nature, its adherence to excellence in small matters as well as large, its traditional concern for how human and machine—whether sword or teacup—fit together. It would—and this was a key point—be named as something separate from the usual Nissan line of cars, and be sold and serviced in a separate and different way from the regular line.

This was a somewhat controversial recommendation. All new ideas are, but this one was especially so, for to celebrate a Japanese luxury car had never been done before. (It was definitely not the safe recommendation, which would have been to position the new car as simply an expensive Nissan; we thought that such a line extension would be the worst idea, for it is always difficult to sell an expensive new product under a master brand that connotes something inexpensive. Would you buy a sirloin steak from McDonald's?)

Nissan accepted the rationale for the positioning, and we moved on to the next stage, arriving at a name for the car. Here were some of the criteria:

1) It must project the notions of luxury and quality; be consistent with the idea of an international, "world-class" product; and connote state-of-the-art technology. Top of the line. None better.

2) It must nourish the idea of exclusivity—prestige—status—and signal all of these to discerning, differentiating car enthusiasts.

3) It must project the image of uniqueness and of those timeless/excellence attributes that we associate with the best of Japanese culture, as well as the personal enhancement to be gained from purchasing the product and having that purchase appreciated within one's social set.

4) It must project the image of a product keyed to provide customer satisfaction, beginning with the "human engineering" design of it, and carried through in the way it is sold, in personalized owner care, and in hassle-free service.

5) Beyond these, the name would have to be proprietary—not a dictionary word, but sounding as if it were, and conveying the idea that the product so named would be the ultimate car you could buy.

Of course by now you have long since figured out that the name we recommended was Infiniti, so let me make an aside about this name. In our experience, when a board or a management committee is presented with a name, it never happens that people instantly utter "Eureka" and go home satisfied—but with the recommendation of Infiniti, we came as close as I think we'll ever get to that blissful state of affairs. Everyone agreed that it met the criteria we had established.

The next step was the logo, and that was easier once the name had been settled. The ancient symbol for infinity, the closed loop that looks like a figure eight on its side, immediately came to mind, and our design was a modification of that. In today's global marketplace, a symbol must gain two kinds of approval: first, it has to meet the manufacturer's needs, and second, it must be meaningful to the company's customers. In this, we were a bit lucky, for when we showed the design to Nissan, the Japanese associated the inner angular peak in the center of the oval with the shape of Mount Fuji, the most famous—and sacred—symbol to the Japanese. This might have been buried somewhere in the back of our minds—but our research showed us that when American consumers saw the design in conjunction with the Infiniti name, they thought the inner angular configuration indicated a road receding ahead of them into infinity. In any event, it gained the approval of both groups, and was that happy marriage of symbol and name that we always strive to create and only do so when the gods smile on us.

Then came the hard part. Our goal, and that agreed to by Nissan, required more than a name, a logo, and its associated design system. To fulfill the objectives we had set out—that is, to have a luxury car that was fully differentiated from the other Nissan products—the Infiniti would have to be introduced, sold, and serviced through an entirely new distribution channel, and not in tandem with other Nissan cars, trucks, sports cars, vans, etc. The decision was made that it be sold only by dealers who had a separate showroom in a building physically distinct from its other showrooms. To make such a "second channel" would be extremely expensive; the individual dealer might have to allocate two or three million dollars to build and maintain the sort of separate facilities we thought essential to the concept. In effect, Infiniti was to become a separate division of Nissan, in the way that Cadillac is to General Motors. The Infiniti identity would meet the needs of both the brand and the division.

The difficulty was that because the car itself was not yet available to be seen, photographed, and described, it would be necessary to convince the dealers to go ahead with an extremely expensive proposition on the strength of the image Infiniti hoped to possess and project. We needed to find ways of demonstrating to the prospective dealer that Nissan was putting its money where its mouth was, and would justify the dealer's investment by spending hundreds of millions of dollars of its own. We designed a series of communications tools to convey these ideas to potential dealers. No expense was spared in this effort. There were prototypes of dealer signs, designs for showrooms, books and manuals on showroom communications—an entire package to convince dealers that they would have a luxurious and very special product to sell, and resources from the parent company that would combine to make the car attractive to buyers. We even built a model of an outdoor sculpted signpost for Infiniti, lit it to show how striking it would appear at night on the site of a dealership, and included photographs of that.

We designed "corporate media" for the dealerships and showrooms that adhered to the idea that Infiniti was a separate operation; on business cards, forms, stationery, one read only the name Infiniti and the logo.

Expectations on the customer side were reinforced by an intriguing series of advertisements done by an agency engaged by Nissan USA, ads that raised eyebrows because they didn't show the actual product. All they were selling, at that point, was an image. We did not participate in making these terrific ads, but were excited to note that their essence was a restatement of the basic positioning concepts we had recommended at the beginning of the project: they stressed Infiniti's Japanese character, the traditional Japanese relationship to nature and quality, and invited the reader and viewer to equate these with a new understanding of luxury.

When the Infiniti was finally introduced at an automobile show, everything that surrounded it was distinctively Japanese, from bamboo frames for the background to the foliage used in the foreground. This, too, was a reinforcement of the message Nissan wanted to convey. A logo, when put on a car, is called a badge. Since we hadn't seen the actual car when we'd made the logo, we were now pleased to see that it stood the test of good badges, that it could be applied in both expected and unexpected ways to various parts of the car from the hubcap to the dashboard, and still maintain its integrity. Infiniti's logo also appears on the keys to the car, which is yet another element buttressing the equation between the product and luxury—the key becomes a key to luxury.

Infiniti was introduced during the fall of 1989, and its initial awareness level was at an unprecedented high. As a measure of the detail to which Nissan committed itself to the positioning and design theme, by way of greeting and thanks, Nissan sent to dealers a distinctive and expensively original origami Infiniti Christmas card, the final touch in a fully integrated identity and image management program.

CHAPTER 9

Pluses and Minuses

As pleasant as the NYNEX and Infiniti programs were, others, while equally as difficult, were a lot less smooth . . . And I mean a lot! One of my worst moments came when Donald Trump suggested publicly that Allegis was a disease from which a certain airline was suffering, a disease that had no cure.

The Allegis Debacle

In the early 1980s United Airlines formed a holding company known as UAL, Inc., for the purpose of buying a hotel chain. The airline industry was regulated and prohibited from investing in anything other than its airlines—but the green-eyeshade guys figured that since they were spending so much money on lodging for cockpit personnel and flight attendants who were away from their home bases, owning their own hotels would mean big savings. The chain in question was Western Hotels, actually a former client of ours to whom we had recommended a name change, to Westin, to reflect its planned global reach.

Later on, UAL bought Hertz Rent-a-Car, purchased more hotels from Hilton International, and developed the Apollo computer reservation system as a profitable service for all travel agents. Clearly, the parent company was considerably more than an airline.

In early 1985, UAL CEO Dick Ferris called us to a

meeting to talk about changing the parent-company name. Ferris thought this would not be a big deal, that it could be done in a month, and that he would introduce the name casually, in the company's in-flight magazine. On further questioning, we learned that the major reason to consider a name change was that UAL shares were selling at about six times earnings—a ratio that held for most of the other stocks in the airline industry at that time. Ferris wanted to change the name in order to help the financial community recognize the company's true position as a purveyor of more than airline seats, and to induce more than only airline industry analysts to follow the stock; if more people were aware of UAL's size, particular combination of premier travel services, and mission, he was convinced that the stock price would rise. He believed that changing the corporate name would be a simple process, and one for which he could easily obtain approval from his board of directors.

My associates and I told him a horror story about another large company and its CEO, the legendary Harry Gray, whose impeccable business logic convinced him that he would get a rubber-stamp go-ahead for a name change from what was then named the United Aircraft Company, but who instead ran into an embarrassing, costly, and distracting buzz-saw of resistance because he underestimated the amount of emotionality attached to a company's formal name, and the stress associated with altering it. I reminded Ferris that a corporate name change also required shareholder approval, a particularly high-visibility process.

On the spot, Ferris telephoned several key board members such as astronaut Neil Armstrong, Charles Luce (former chairman of Consolidated Edison), and former U.S. treasurer Juanita Krebs. He set up meetings for us with them and delayed proceeding until we could report to him on their feelings about a change. After obtaining their approval for the conceptual direction, we were given a go-ahead to start our work, and a year in which to do our job properly.

Our interviewing and other research produced the conclusion that so long as the parent company was known as UAL, it would be unable to divorce its identity from that of United Airlines. UAL was not only the name of the parent, it was also the stock symbol and the acronym that appeared in countless print and television advertisements, and in signage all over the country; it was even placed on United's baggage tags. UAL was a ubiquitous symbol for the airline. In the future, the company hoped to increase revenues from non-airline assets, and to become less dependent on the fluctuating profits from air travel. It was clear that a new identity for the parent was mandated in order to achieve Dick Ferris's objective for the stock. However, we saw any potential name change as being aimed solely at changing the parent company's image in the eyes of the financial community. Any other potential uses of the parent identity were, at best, off in the wild blue yonder. At no time was a replacement identity even remotely considered for any of the enormously valuable trade names owned by the parent, such as United Airlines, Hertz, Westin, or Hilton International.

Our leading contender for a new name was Allegis. It had as its roots the English word *allegiance,* a word derived from the Greek *egis,* or sponsor. Allegis connoted something that was faithful and loyal, caring and protecting. We hoped it was appropriate for a parent company whose divisions served the traveling public. It was distinctive, memorable, and deliberately disconnected from the airline business; it was a clean slate to be used to encourage the financial community to focus its attention on the new reality of the organization.

Candidates for new names were reviewed by the management, and then by the board of directors, which approved the name Allegis.

All well and good so far. But then Dick Ferris conceived the notion that this name was far more than a parent company name. His idea was that all of the company's businesses—airline, reservation system, hotels, and rental cars—served the same customer, the affluent businessperson, and that in the future all of these services could be integrated. Allegis would be able to boast that it could take a customer (and his bags) from a Westin Hotel in one city, through a Hertz car trip, onto a reserved seat on United, into another Hertz car in a second city, and to a Hilton International Hotel in that second city—all as a result of one telephone call, with a minimum of paperwork at the various counters, and without that customer's bags having been misplaced or delayed. Allegis couldn't yet do this, Ferris said, but it could be accomplished in the near future. In short, he saw the Allegis identity not simply as a parent company name, but as a marketing tool. And he determined to introduce the name with a big splash, as the coming miracle in travel services.

We at Lippincott & Margulies were stunned, because we believed that an identity fashioned for the purpose of communicating a parent's reality to the financial community must be introduced in a direct and appropriate way to its specific audience and should never be positioned as a marketing

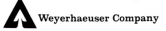

Coca-Cola

The Coca-Cola Company had a problem. Their world-famous flagship brand was being presented in many shapes, colors, and design styles—undisciplined expressions that diluted the impact of their vast promotional activities.

Needed: a design system that would be highly visible, memorable, and instantly recognizable in any language, from any distance, and on media as small as a uniform badge or as large as a truck.

From the company's history we isolated four key identity elements: the Coca-Cola name in its historic script logotype, the color red, the word Coke in its traditional typeface, and the shape of the original Coca-Cola bottle. These became the foundation of the coordinated design system that today conveys the identity of brand Coca-Cola worldwide.

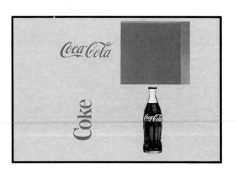

Chrysler

What was Chrysler the brand? How was it different from Chrysler the corporation? Did a Chrysler dealer sell only the Chrysler brand or all brands made by Chrysler? To help sort out the confusion, we invented a symbol, the pentastar, to identify all things made by Chrysler, and a design system to boost the appeal of the different brands. After the "new look" was established, sales jumped. Iacocca liked the pentastar so much he found even more uses for it.

Humana

Extendicare was an adequate name for a chain of nursing homes, but when the company decided to exit that industry and enter the hospital field, a new identity was clearly needed. As a name, Humana preempted the idea of a health care company dedicated to "humanitarian" caring for patients. A design system was developed to raise the awareness level of this new identity. One example was linking the Humana name with the original names of the hospital it acquired, as in the Humana Hospital of Louisville, the site of the first artificial heart operation.

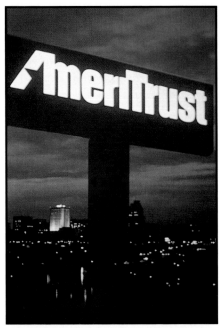

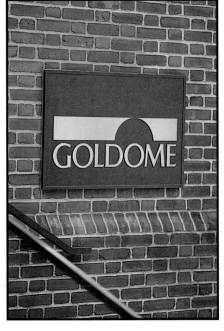

Hundreds of companies, especially banks, have changed their names to fit new business strategies, support marketing efforts in new geographic regions, or to understand better the pressures of new competitors. Most customers forget the old names very quickly. Did you?

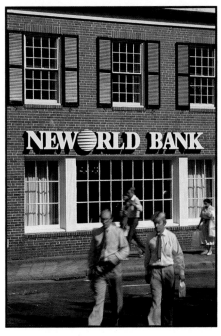

Infiniti

Task: Give us an image promising the ultimate in luxury; technological excellence. Creating the Infiniti name was our first step. Next: to evolve from the symbol of infinity a logo for the automobile that would be distinctive, elegant—and proprietary.

Kellogg's
The Kellogg's Corn Flakes package through the years, from brand name only to appealing imagery messages for the times: kid's stuff, healthful food, to just plain delicious.

Amtrak

Starting something from scratch is always fun. Congress had decreed that railroads could divest their passenger services, which were to be unified and positioned as a modern, up-to-date form of mass transportation. For the new national passenger railroad system, we created the name Amtrak. We designed every item that a passenger or passerby could encounter, from cocktail napkins to tickets to the look of the trains both inside and out. Today, most Americans equate the name and the familiar red, white, and blue design scheme with railroad passenger service.

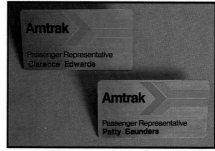

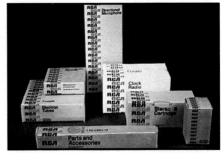

RCA
Leading America into the electronic age. From Radio Corporation of America, with its original logos featuring a bolt of lightning or Nipper listening to a wind-up phonograph, to RCA, with a totally modern, truly international logo, designed to "read" in more than one language and not be limited by any single electronic product.

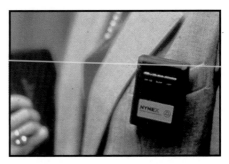

NYNEX

Challenge: Establish a completely
new identity for a brand new
company which at birth has $16
billion in revenue and 110,000
employees. Build awareness for
this unknown company across a
vast array of media, including
everyday stationery, vehicles,
building signs, packaging, product identifiers, and advertising.

 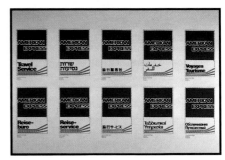

American Express
American Express asked us if
their name should be changed.
Keep the name, we said! Increase
awareness of it by adopting a design system and identity practices
that will enable customers, investors, and the general public to
appreciate the many different
financial, travel, and insurance
services American Express offers
under a single umbrella.

Duracell

To knock the market leader from its perch, the Mallory Battery Company wanted a new identity for a good product—a new name, new product design, new package design. It had to connote superiority, durability, and difference from all others. Now, everybody knows and recognizes the copper-topped, black-bottomed Duracell, which has become the undisputed market leader.

Primerica

American Can had sold its can-
making division and name and
became a conglomerate, encompassing Sam Goody's, Fingerhut,
Smith Barney, and other financial
services companies. They now
did not own the old name—and
anyhow it no longer reflected the
company's reality. This cartoon
commissioned by *The New York
Times* captured the "dozens of
meetings, hundreds of man-
hours, millions of dollars and
months of angst" that went into
the name change.

concept, even if the marketing aspect was a secondary consideration. Then, too, Allegis had never been designed with the general public in mind; it was just for the small, highly influential financial community. Furthermore, we were concerned because Ferris's claim wasn't credible: consumers wouldn't believe that they and their baggage could be magically whisked from one place to another without a hitch occurring somewhere.

In fact, back at that first meeting with Ferris, he had told us a joke originally attributed to Bob Hope. One day Hope had arrived at the Los Angeles Airport with five bags and instructed the skycap to send one to New York, one to London, one to Rome, one to Frankfurt, and one to Chicago. The skycap informed the comedian that his request was impossible and simply couldn't be accomplished. Hope asked why they couldn't do it this time, since that was exactly what they had done by accident on his previous trip. At the meeting with Ferris everyone laughed, as I expect audiences everywhere did when Hope told the joke, because Hope's attitude was widespread: people recognized the general atmosphere of confusion over complicated baggage transfers. It would be flying in the face of this attitude to introduce Allegis not only as a corporate name but also as a travel concept. While the level of efficiency made possible by the advanced computers of the Apollo reservation system was a giant leap forward, there was just no credibility for Allegis to function as Ferris promised.

Nonetheless, the Allegis idea was duly launched with a huge splash, and before a large audience garnered through advertisements in many different media. Chaos ensued. Pundits like Trump dubbed Allegis as the name of a new disease. No convincing effort was mounted to differentiate the parent from the airline. People thought the airline was being renamed because the marketing concept was provocative—and unbelievable. Although the awareness level for Allegis went through the roof overnight, the concept was ridiculed, and the symbol for the concept—the name—received the brunt of the derision. Earlier in the decade, Ferris had had a difficult time with the pilots' union; he had won a contract dispute, but the victory had left a bitter taste in the pilots' mouths, and they took the occasion of the introduction of the Allegis identity to issue a public outcry suggesting that the name change was really a reorganization of the company that would injure the health of the airline. Dick Ferris thus became the man whose strategy provided his enemies a tool with which they could attack him.

Perhaps the most damage done by the overblown introduction of the name change was in awaking the breakup artists, who now did their homework and readily concluded that what had been UAL, Inc., would be worth more when broken up into pieces than if it remained an intact conglomerate. At $45 a share, they agreed, Allegis was quite undervalued. This, indeed, had been Ferris's contention—but now his company was in play. This was when Trump made his "joke," and simultaneously offered $75 a share for the "disease." Ferris attempted to make a deal with Boeing to offer the equivalent amount, and Trump withdrew. Another bidder weighed in at $90 a share, and Ferris was unable to match it.

Astonished at the uproar, the board now forced Ferris to resign, agreed with the predators that it was in the best interests of the shareholders to break up the company into its separate pieces, and rescinded its earlier approval of Allegis, changing the name back to UAL, Inc., because, indeed, there was no need for a separate parent structure and attendant identity. All the company's non-airline assets were put up for sale. Many of them brought far more than expected. The Ford Motor Company backed an LBO of Hertz by its employees. The Bass brothers bought the Westin Hotels for one billion dollars and subsequently sold just one of them—the Plaza, in New York—to Donald Trump for $400 million. The Apollo reservation system, which we renamed Covia, in order to position it as more than an airline reservation system, also sold for $500 million.

These events were a bonanza for the stockholders of UAL, Inc., who received cash for the divestitures of these various parts; in addition, the stock price of UAL, Inc., soared above $100.

In our view, the operation was a success but the patient died.

The name Allegis became a symbol for the debacle that happened to the organization. We once again drew a lesson for ourselves—many other instances had previously contributed to our understanding of it: that the problems we address always emerge from developing circumstances in a client company's business. A name change cannot be a superficial attempt at image manipulation; rather, it is a serious matter, for the name is the key ingredient in the psyche of the company. Altering it means a reorientation in how the employees think about their employer, as well as in how the public views the company. The name is the capstone of an identity program; a corporation's image is the result of a process of growth and development; the name must not seem to be a gloss applied over the top of an entity to hide its inadequacies—it must be reflective of the realities underneath. How a name is introduced, the reasons for it, and its context must be as carefully considered and controlled by a company as is the thinking that leads up to the decision to change a name or institute a new one. Essentially, a name is an abstract concept capable of evolving into whatever management desires it to become, but care must be taken so that it will succeed, a success that is always measured in terms of how the name and the corporate mission it reflects are understood by the corporation's crucial audiences.

The matter of an identity being targeted for a specific audience was the uppermost problem in the generation of another name for which Lippincott & Margulies has been both praised and blamed, Primerica.

One Company, Changing Identities

Consider the odyssey of the American Can Company. The corporation that bore this name had long been considered the quintessential blue chip of American corporate aristocracy. For instance, American Can had been the fifth stock to be listed in the Dow Jones Industrials averages. Starting in 1970, the company began to diversify, and bought such things as a specialty retailer, Fingerhut, some financial services concerns, and the record-and-tape retailer Sam Goody's. Then financier Gerry Tsai sold his own financial services company to American Can, and, after a series of maneuvers, became CEO of the parent. By 1981, American Can had changed from a company principally involved in packaging to one that was service-based. In 1986 American Can took the extraordinary step of divesting its core packaging entity, selling the name and the American Can division to Triangle Industries. (I'll say more about this transaction just below; for the moment, though, let's stick with the parent.) Having sold American Can, Tsai needed a new corporate name, for two reasons: first, because he no longer owned the old name since he had sold it along with the division, and second, because the remaining business had to be understood for what it had become, not what it had been in the past.

Underestimating both the significance and the difficulty of the task, and rather than hiring Lippincott & Margulies right away, Tsai first asked his executive committee to come up with a name, offering as an incentive the munificent sum of five hundred dollars. Tsai received a stack of suggestions whose three-foot-high density matched their inability to communicate well. Only then did he throw up his hands and dial our number.

Now a company's name is not something isolated from that company's purpose; it is only the capstone of an identity program. So in our first meeting I asked Tsai one of our standard questions: "What business are you in?" That turned out to have been the crux of the matter, for Tsai felt he was not in any particular manufacturing or service business, but, rather, was in the business of maximizing returns to stockholders—period. In what ways that was to be accomplished, and through what sorts of industries, was not important to him. Therefore he required a name that would enable him to buy and sell any company at any time he wished. Unable to predict what the company's main business might be in five or ten years, he felt, frankly, that this was a strength, because it allowed him not to be tied down to any particular industry or line of business.

Fairly irregular, you might say, but it did not constitute an impossible request. Our job was to fashion something that would work for Tsai and help him move toward his objectives. We concluded that the new name need only be appropriate for a parent company, and, perforce, we ignored product and service marketing considerations of that new name. We did believe, however, that once it had been chosen, he should preclude any subsidiary from using it. The reason for this was complicated but understandable: if the subsidiary had a name other than that of the parent, and it was later sold, Tsai would then easily be able to retain the parent name. If he used the parent name also for a subsidiary, and later wanted to sell the subsidiary, a prospective buyer might offer less for it if he couldn't purchase the established name along with the manufacturing or service unit.

Another requirement was that the name be totally non-restrictive, both as to geography and as to type of industry. A further restraint was that Tsai already knew what name we ought to propose. He had convinced himself that names ending with an X, such as NYNEX, were favored by investors and invariably represented companies with better P/E ratios than their competition—and so his candidate for a new name was XAMAX. Fortunately, with the help of his former chairman Bill Woodside, we were able to dampen his enthusiasm for this label. Eventually we came up with a name that was both a bridge to the company's former identity and a link to its financial base; it combined the ideas of "prime" and "America" into Primerica.

We liked Primerica, Tsai and his board liked it, and after approval by the shareholders, the name was introduced to the financial community in a series of advertisements featuring actor John Wayne (whose own name had been changed, long ago)

Image by Design

and the slogan, "Sometimes the name you grew up with doesn't fit you anymore."

It would be nice to report that the cowboy then rode off into the sunset and the movie ended happily, but, for better or for worse, that wasn't the end of the story.

In September 1987 Tsai bought the brokerage firm of Smith Barney for two-and-a-half times its book value, the highest price and multiple ever paid for that sort of Wall Street house. A month later, stocks and bonds experienced the crash of '87, damaging Smith Barney severely enough to make an impact on Primerica's financial well-being. The asset value of Tsai's company became so depressed that Sandy Weill's Commercial Credit Company bought Primerica. To cap off the long saga, Weill did something most acquirers never do—he gave up the well-established name of his old company in favor of Primerica. Why would he take on the name of the vanquished company? He did so in order to retain the position of successor to American Can in the Dow Jones averages, and in order to benefit from the superior imagery that had already become associated with the Primerica name. Within record time, the Primerica identity has become institutionalized, and, in what insiders know to be a truly remarkable feat, is now a well-respected blue chip identity.

Details and Distinctions

CHAPTER 10

Logic, Creativity, and Sensitivity

Image-related alterations help solve mainstream business problems. But how does image making work? Here is how we go about trying to ascertain a business's needs and future expectations, and design communications solutions to help implement them. The key factors are logic, creativity, and sensitivity.

Questions Come First

When Lippincott & Margulies begins a program for a corporate client, our initial task is to identify and clarify a company's strengths and weaknesses through interviews with management and with selected external audiences. These audiences may include financial analysts, competitors, customers, consultants, government and community leaders, and the business press. Our aim in these interviews is to determine how management views itself, the company, and the company's future directions and goals, and to contrast these views with how outside audiences see the company. By identifying and measuring the gap between these perceptions, we come to some hypotheses, and then design a communications system and create identity tools with which the company can close the gap.

Interviewing is the key to the whole process. When we make our recommendations, we like to be able to cite as evidence for our conclusions the information we've gathered, replete with "site-specific" quotations. Part of the reason we do so is to avoid making the impression that our suggestions are merely the product of our own minds. Most often they come directly out of what we have heard from internal and external interviews, and the patterns we've discerned from them. Over the years, my associates and I have conducted thousands of interviews with executives in large companies, and have evolved some general understanding and principles for having successful—that is to say, useful—interviews.

Internal Interviews

Internal interviews are the most useful ones, since it is the company's management that determines how it wishes to be perceived, knows best which business strategies it will adopt for the future, understands what identity practices are likely to be the most credible within the context of the firm's own history, and realizes what positioning is culturally acceptable to its own employees. It is the company's management that is ultimately responsible for implementing any of our recommendations; management alone will control the company's identity elements.

We have some standard understandings for such obviously critical internal interviews.

When we are hired by a company, it is with and through the consent of the CEO or chairman. Having set out the field of endeavor, we then leave these top officers for some time, and interview many other people in the company, anywhere from a dozen to a hundred different people from all categories of management. If it is a corporate identity program we're developing, we see a variety of managers; if it is a specific brand, we'll logically concentrate more on those people who have direct responsibility for that brand.

It is essential to the success of any interview that the interviewee be informed by his or her superior of the subject, of its importance to the company, and of the fact that it is being conducted at the behest of the company's top management. It's always a good idea for these matters to be broached with the interviewee ahead of time, so that he or she has a chance to think about the subjects and formulate ideas. We are, after all, delving into areas that touch upon the heart of what the company is, does, and hopes to become, and it is of assistance to have interviewees reflect a bit on such subjects if they are to give us truly helpful information. And let me repeat: that's our goal in this stage of the process, the obtaining of information.

We generally are given a list of executives to be interviewed. Such lists invariably contain two sorts of names—those managers who can increase our knowledge of the company's history, status, extent of its problem areas, and future strategic plans, and those who need to be interviewed because if they were excluded from the process they might feel slighted and inclined to work against the recommendations. While we are not told which category any given interviewee fits, the reasons for their inclusion usually become apparent during the interview.

We find, too, that each interviewee has two agendas. One is visible, the other, hidden. The first has to do with the stated objectives of our interview, the subject of the company's history, problems, future, and so on. The second agenda often deals with what actions the executive would like to see taken as a consequence of the corporate identity program, and the executive's attempt to understand and influence how this program can be used for his or her own benefit. A skilled interviewer can also separate these agendas during the course of an interview.

I must point out parenthetically that this second agenda is neither unexpected nor unwelcome, for it is essential to the success of the entire enterprise that all executives feel themselves a part of the recommendations and feel they are able to reap something for themselves from them. Part of our work in conducting so many interviews is building a consensus for the corporate identity program itself.

A principle that we adhere to religiously is to conduct all interviews on a one-to-one basis, to announce that it is confidential, and to tell people that we may include their conclusions in our report, but never in such a way that they are attributable to specific individuals. We do this in order to have frank and full discussions, and we back up our promise by not using any memory aids exept pencil and notepaper.

We work our way up the ladder, back toward the CEO and chairman, whom we interview last of all, so as not to waste their time obtaining the sorts of information we can get from lower-ranked executives. As we work through the interviewing process, we try not to formulate hypotheses about the company's problems or possible solutions; we look for patterns in what people tell us, and try not to rely on a single comment from any single individual. If twenty people all let you know, in different ways, that the company's problem is a lack of morale, such a pattern cannot be ignored. Inevitably, though, as we go along we do begin to recognize these patterns and to ask questions based on them and on possible solutions as we come closer to formulating them. What we hope we're doing is querying a company's executives in a way that hasn't previously been done, and in the process discerning their problems and analyzing them from a perspective that will yield optimum solutions to image-related issues.

External Interviews

These are conducted in the same confidential way as are internal interviews, and our goal in them is the same: to learn how the interviewee views the particular company. We try to prepare extensively for such external interviews, so as not to waste the time of the executives we are questioning. There are several reasons why outsiders would sit for such interviews, chief among them a wish to please the client company, and, not far behind, the unequivocal fact that everyone likes to be asked his or her opinion and to have those views assiduously recorded. We find more often than not that the higher the level of an executive being interviewed, the more professional his or her attitude, and the better the information we obtain.

Perhaps the shortest—and, I should say, the most prescient—"external interview" of this sort was one that I held with a well-known mogul who agreed to see me in connection with an image study we were conducting for a prominent company in the investment community. I told him the client's name and that we wanted to find out how the company was regarded by its peers. His opening question to *me* was, "What price/earning multiple does your client's stock have on the Big Board?"

"Seventeen," I replied.

"Then what are you doing here?"

To translate: seventeen is a particularly high and desirable multiple, and meant to him (and was supposed to mean to me) that a company that had such a nice multiple could be said *ipso facto* to have no image problems. While his answer had value to us as we considered how much of a change to recommend to our client, it also demonstrated in an unmistakable way the quality of thinking that is often available to us as we strive to fashion a recommendation.

Since Lippincott & Margulies is by now well known as a consultant to large and important clients, external executives are disposed to cooperate with us, which helps the process along. Most of the time, in these external interviews, we are able to tell people the precise company with whom we're working, but sometimes not; in the latter instances, we couch our questions in terms of an industry, and may include the target company within a list of others and obtain opinions on all of them. While the latter condition imposes a limitation on the quality and quantity of the insights we can gain into the client's circumstances, it is sometimes necessary to suffer these because confidentiality is of utmost importance—for instance, if there is a fear that the consideration of a change in identity might leak out prematurely and be misinterpreted by crucial audiences, competitors, or even the SEC.

The most useful external interviews are with customers, competitors, and the business press, and the most difficult are with financial analysts. The latter always protest that they don't care about a company's image—"Just give us the numbers, please, and we'll do our own interpreting of them"—but prove to be the biggest gossips, seizing on any rumor in their attempt to get the *real* understanding of the direction and strength of the subject company.

To Be a Good Interviewer

To be a good corporate identity interviewer is not a simple task. Chief among the characteristics I deem important is the ability to ask a series of questions within a business context that is appropriate to the situation—not in a narrower definition (such as the design or name concerns), not in a personal sense (whom do you like, what don't you like), but in a context that is thoroughly professional, broad-based, and oriented toward long-range objectives. This includes asking questions that are appropriate to the interviewee, that clearly relate to the subject at hand—perception of the company—and about which the interviewee can be expected to give valid opinions. To overlook the requirement for concentrating on how the company is perceived is to risk raising the suspicions of the interviewee about the true purpose of the interview and of the program itself. The interviewer must make it crystal clear to the interviewee that he or she is in no way being grilled on how well or badly he or she is doing.

Good interviewers do not constrain their observations to words alone. They are always alert to body language, nuances, inflections, raised eyebrows, sighs, smirks. A sigh coupled with a phrase such as "If our honorable chairman ever makes up his mind" can provide as valuable an insight into the reality of a corporation as a well-reasoned response to a carefully crafted question. While a degree in psychology is not a prerequisite for being a good interviewer, a thorough understanding of the dynamics of the executive suite, and a knowledge of how people respond within that context—a mix of personal ambition, executive values, and leadership styles, to name just a few of the elements—are vital to the process of accurately weighing and analyzing the information being gathered.

It's important for all interviewers to inject a little humor into the situation; we do deal with serious subjects, but an interview without humor is apt to lack a certain down-to-earth quality that seems to coexist with realistic insight into the company's problems. We can't guarantee that our interviewees will have that quality, but we can insist that those asking the questions do so in such a manner as to heighten the probability of finding it.

The good interviewer also tries to strike a balance between the inside knowledge he or she is gaining as a result of a series of interviews, and the outside posture that allows us to elicit information and opinions that are different from those that might be obtained by a complete insider or given to a complete outsider.

Good interviewers learn to treat themselves as the equal in status of the person being interviewed; they do so without arrogance, without undue humility, without looking either up to or down on the person answering the questions and offering the opinions. We make it a rule never to interrupt when an answer is being given, even if it's very wide of the mark; to do otherwise would be discourteous.

The knowledge gained from these interviews provides the tools with which we can begin to solve the problem at hand.

The Design System

After we gather information from interviews and analyze it, we come up with recommendations for comprehensive communications practices that will address the corporate positionings that have been selected. Our solutions are generally expressed in two modes, the visual and the verbal; or, as we'd put it in technical terms, the design system, the name, and the nomenclature architecture.

Companies are continually sending out visual and verbal messages to change or to strengthen their image in the marketplace. Eighty-five percent of what we learn is perceived through our eyes; therefore, the visual element of any image is extremely important. Design components of an identity are the things that last; long after the effect of rationalizations and other thoughts (often expressed in words) has dissipated, the visual elements remain. Design systems can represent the single most significant capital expenditure associated with the implementation of a new identity project. Within design systems there are both permanent and ever-changing components. The permanent ones are considered the traditional mainstays of identity—stationery, forms, signage, motor vehicles. These lend themselves to being tightly controlled for consistency and to combine to produce the maximum impact. But there are changing items, too, principally sales and promotional materials, which must be capable of being modified rapidly in response to changing market conditions.

When we investigate a company's "communications architecture," it is to find out such things as the cumulative effect of all the company's brochures. What message are these sending? What do the company's packages communicate? Is there vertical integration in the company's presentations to its various audiences? The design system, then, is more than just instructions of how and when to use a logo. *The design system is a device enabling the organization of all visual expressions emanating from a company in an economically favorable manner to produce maximum impact.*

Two Uses for Design Systems

How do the visual expressions of a company contribute to—or hinder—the company's strategies? Design systems, the embodiment of the visual elements of an image, can be used by companies as marketing tools and also to provide economies of scale. Most people understand design systems as marketing tools. As such, visual design systems that include logotypes, colors, typefaces, and other elements can make the small look large, or make all parts of a company appear the same. For instance, the look of IBM is the same throughout all of its subsidiaries. That sends out a particular message about how IBM is structured, and how its strength is reflected in its component divisions. The look of Procter & Gamble in the marketplace is entirely different: each product and each division appears distinct from the others—you would hardly know from glancing at the packaging that Tide and Pampers are both P&G products. In between these two extremes is the design system of United Technologies, wherein the parent company is strongly associated with the subsidiaries, but each subsidiary has its own look.

Design systems can also be used to provide companies with economies of scale. Quite often, we are faced with a challenge from a client: the company is spending $10 million per year on signage, but that money is not buying a visual look that strengthens the company. In our investigations, we usually discover that the signage is being ordered by individual managers, many of whom are in locations remote from headquarters, and who are paying premiums to local sign-makers to design and deliver the needed materials. A similar ordering scheme exists for business forms and other stationery—locally ordered, their cost includes buying the design of the forms at that location, and paying extra for creating that design.

If we can help the company clearly articulate its policies and practices in the visual design area, we are on the right track, since a design system that represents the headquarters' thinking can then be passed on to individual managers and used at their locations. Or, signage, business forms, and other stationery can be managed centrally and sent out as needed to the remote locations. In general, when central control is exerted, companies can achieve significant annual savings on the cost of signage and other expressions of their identity.

In old-style corporate identity programs, the design group and headquarters would create an identity and then stamp it across all the company's communications and products. Today, we find that this is too rigid a notion for some companies. Now, especially in regard to what I've defined earlier as the "flexible components," we identify the elements of the design that headquarters wishes to emphasize, and specify a particular family of typefaces rather than a single type, and a family of colors rather than a single hue; we issue a stylebook that allows individual managers to pick something from the family of type-faces or colors. If these managers then go to local suppliers, they can at least order their signage and forms with design elements already specified, thus lowering the costs, though not as dramatically as when these visual expressions are bought centrally.

Our role as design experts is not only to create appropriate visual expressions of the company's identity, but to build consensus around the range of standards that we propose—for if management is not enthusiastic about them, these expressions may not fulfill the hopes of the company's leaders. Lippincott & Margulies often takes the role of arbiter, and does as much work in getting the standards accepted as we do in finding appropriate designs in the first place.

Three Design-Driven Systems

Ryder Systems once had to buy back the exclusive rights to its name for all its trucking activities. The formerly identified Ryder/PIE fleet then needed a new design system. We identified its equity elements as the name Ryder and the color yellow, and convinced the company that these elements had to dominate any visual presentation. We also found a red line that ran under the Ryder logo in some of the old expressions. After looking further into this, we decided to run the red line entirely around the truck at a certain height above the road, not just for aesthetic reasons but primarily for reasons of safety. Liking the idea, the client asked its insurance carrier for a reduction in premium rates because the company now boldly exhibited its distinctive safety aid. The rate reduction was approved, which generated considerable savings for the company.

Unfortunately for us, the division carrying this design format was later sold at the time Ryder again decided to leave this section of the trucking industry, and so the design system can no longer be seen on the roads.

The Rustler chain of "steak houses" was becoming a bit outmoded because customers were being admonished by doctors and the media to eat less red meat and barbecued foods, and more salads, granola, and other healthful concoctions. Previously, the architecture of most Rustler restaurants had been designed to look like an old Western town; the fronts of some of the buildings actually looked like a series of buildings, and that caused some customers to have difficulty in finding the proper front door. The restaurant layout itself centered around the grills for the meat—these were, after all, steak houses—and it was the grills people remembered and that some customers were beginning to avoid. Since we understood this, our recommendations went beyond signs and corporate communications to the actual design of the restaurants themselves, inside and out.

We recommended a change in nearly every element of the Rustler restaurants. The idea was to convey that the restaurant was a healthful environment consistent with the life-styles of customers aged twenty-five to forty-five and their children. Outside, where possible, there were lots of shrubs and trees, and the name Rustler was rendered in green rather than red; inside, there were greenhouses and skylights. The salad bar was the most prominent feature, and the grills were secondary. This was so upsetting to one executive that he uttered a classic short-order critique: "What's with you guys? I'm trying to sell meat and now everybody's eating grass!"

It was a terrific program, and we feel it would have transformed the chain from steak-and-barbecue houses to family restaurants appropriate for today's changed emphasis in eating habits. Unfortunately for us, though not for the shareholders, six months after we'd built two prototype restaurants, the chain was sold to Sizzler Steak Houses, and the pilot program languished.

The chain of Loeb food distribution warehouses in Canada approached Lippincott & Margulies a few years ago in a characteristic way, wanting a new logo; the CEO didn't like the way it looked. As we investigated, it turned out that Loeb's problem was more complex than the desire for a new logo. The company was going to change its direction; while remaining a food distributor, it was now about to enter the retail food business. This entry had come about as a result of having an opportunity to buy at a good price some existing retail outlets identified as IGA stores. However, IGA stores were positioned as average-price-range shops within their category, and Loeb wanted to compete in the higher-priced sector as well. If it was clear to the company that the old Loeb logo wouldn't help, it was as clear to Lippincott & Margulies that the old IGA logo wouldn't be of assistance in this new venture either. So we proposed a modified name and new logo that would be used only on these new Loeb/IGA stores—not on the old IGA stores (some of which Loeb also operated), and not on the Loeb wholesale locations. In the new stores, Loeb/IGA would sell the finest produce, and other better merchandise; we designed identifiers and displays for every element in the store, from the supermarket shelves to the shopping bags.

The concept worked so well that Loeb/IGA became established rather quickly as a leading factor in Canadian supermarkets. Then our very success in helping Loeb/IGA ascend to this level gave rise to a closer relationship between us; we are now working as design consultants for the company's expansion, as catalysts for long-term thinking and strategies. Now we are designing whole new facilities for Loeb/IGA stores, and, beyond that, are helping the company develop branding concepts for new products, such as a Loeb/IGA pizza, and others that can bring additional revenues to the chain.

Some Favorite Design Expressions

Symbols were born to represent a company and its products to the public. Some do this better than others, of course. Here are a few of the ones I like best, principally because they work so well in terms of conveying the proper messages for their companies.

In the gas and oil field, I find the Shell shell a far more memorable symbol than Texaco's star. Hundreds of companies' symbols include stars or suns or rays of one sort or another, but only Shell has been able to take a symbol of petrochemicals— the scallop shell, representing that era in the earth's geologic history when the oil we use today began to form—and make a distinct impression with it on the mind of the public. Shell wasn't so prescient in this regard as it was lucky. Its name was also the name of a dinstinctive shape, and the company was able to bring together the name and the visual symbol in one striking image.

(The technical label for a symbol that resembles or actually is the name of what it is trying to communicate is "iconic"—thus, the shell for Shell, the eagle for Eagle Industries. These began life with more potential to be quickly remembered than their more abstract counterparts.)

Another potent symbol in this field is that used by PetroCanada, a division of the Canadian government: a maple leaf, part of Canada's own symbol, and one that tells consumers that PetroCanada's products are governmentally owned. Phillips has not taken quite the same route, but has built a winning image with Phillips 66. Can you imagine any other number being used as forcefully for a gasoline product?

While we at Lippincott & Margulies are proud of our part in establishing the square block that well represents the integrity of Goldman Sachs, I am an admirer of Merrill Lynch's use of the rampant bull; here, a brokerage house has appropriated a symbol of a rising market, and has used it very well for many years. I doubt that anyone would want a bear, symbol of the market's downside, associated with their investment house; even the Dreyfus lion seems a bit out of character, and I do wish the company would at least get a more vigorous-looking beast to growl in its commercials, as the present one seems so old as to make one wonder if Dreyfus itself has any teeth.

One of my favorite of all corporate symbols is the CBS eye. While being completely unique, it also communicates a specific message. Originally designed and positioned to distinguish television from radio—the latter being the medium of the ear—it has endured and become universally known and respected. Note that this eye is not realistic, for an eye detached from a head could have been revolting; but in its stylized representation it is beautiful and elegant, the perfect visual expression of an identity.

Another excellent example of word and symbol combining to convey a specific message is the cotton mark; again, the design is unique, the message, specific to the thing it is representing. Since obviously relevant symbols are as rare as hen's teeth, the company name frequently offers the best potential. These instances lead me to a basic observation: *Self-reading logos work better than any other, and are the ultimate goal of symbol designers.* Self-reading logos get the message across without requiring the audience to figure out what the logo represents, and without burdening the customer with having to remember and recall separate elements. *The more uncluttered your identifier, the more direct the communication with your audiences, and the more memorable the impression you leave with them.*

Presenting the Recommendation

Before a symbol is given the chance to succeed or fail, it must first be approved by the client, and in this regard the artfulness of the presenter is of great importance. If an abstract symbol were to remind a chairman of a traumatic experience he suffered in childhood, no amount of persuasion will win his approval. On the other hand, even if a recommendation has no such drawbacks, it can be dismissed out of hand if the powers that be become offended by the presentation in other ways. Frequently, these matters balance on the audience's perception of the presenter's attitude, rather than on the terrific qualities of the designs. There is no absolute right or wrong way to make such presentations, just lucky guesses. Many years ago, Walter Margulies—a classic Olympian presenter if ever there was one—was making a presentation for a new logo to the Heublein board of directors, and hadn't done enough homework. He had been warned that Mrs. Heublein herself sat on the board—obviously a person to be treated with great sensitivity and consideration. But there were two women on the board, and at the climax of his presentation, Walter chose the wrong one to charm with the phrase, "And don't you think this is an excellent solution for your company, Mrs. Heublein?" That board member he'd addressed thought Walter had better ask the actual Mrs. Heublein, and it took him some time to dig himself out of that hole.

To come across correctly a presenter must be expert, authoritative, at ease and yet not arrogant.

Now sometimes you can spin a web of professional blarney about a recommendation for a logo or design, when you know that it is fundamentally right—we told United Technologies that there was a precise mathematical formula to describe the symbol we'd developed for the company, even though we hadn't yet come up with that formula—but when a name is at stake, you have entered a realm in which everyone claims nearly equivalent expertise.

CHAPTER 12

What's in a Name? Conflict and Controversy, That's What

Of all the services we provide, the most difficult to accomplish, and to accomplish well, is the creation of new corporate and brand names; I say this even though we have pioneered procedures to anticipate and avoid many of the pitfalls endemic to the naming game. The prime difficulty with naming derives from the fact that the subject is emotional. People form attachments to things (like old names) that are familiar, and they have a fear of the new and unknown. Once, a chairman and his three top aides chose a particular corporate name candidate for their old company, and did so with an unusual amount of enthusiasm. Their company had been known as American Hoist and Derrick, and since they no longer made hoists or derricks, they needed a new name; because they had become a major distributor of a broad range of durable goods, they felt that our recommendation of the name Amdura Corporation was particularly apt. But the chairman's wife didn't like that name, and it became our challenge to persuade her of the soundness of the recommendation; that undertaking was as difficult as anything we'd encountered in coming up with the name. My point is not my temporary annoyance at the lady; rather, I recount the incident to buttress my case that naming is an emotional subject. Everyone feels that they have a valid opinion of a name, and that this opinion is as good as an expert opinion. Would the chairman's wife have ventured a similar opinion about the company's budget for the coming year?

Maybe the emotionality of the subject also explains why corporate name changes seem to stimulate enormous amounts of press coverage, and to get the satirical juices of journalists flowing. Having encountered this so many times, we now alert our name-changing clients in advance that they will have to ride out a storm of controversy—if not outright derision and ridicule—before their new name enters the everyday lexicon and is normally accepted and used in business. Here we're caught between a rock and a hard place. Press coverage is desirable, since it focuses the attention of target audiences on the fact that the company itself has changed, which is the fundamental motivation in making a name change in the first place. Being discussed in the press also results in a rapid rise in awareness levels, something to be desired when introducing a new identity, and it does so at far less cost than an advertising campaign. On the other hand, press coverage of a new name, and the storm it usually entails, produces an unquestionably uncomfortable time for the company, during which many of its competitors take potshots at the company. When the storm clouds of caustic comments start to gather, our recommendation is to have another drink, lie down, and wait until the weather changes.

The Ultimate Accolade

It's hard to believe now, but at the time of its introduction in the 1960s the new name we came up with for the old Cities Service Company—Citgo—occasioned tremendous clamor. The impetus for a name change was evident in the company's reality. Cities Service gas stations were dated-looking, and their green-and-white color scheme faded into the background instead of standing out. Market research indicated that the company's weak visual impact had translated into a lower-octane power image for its gasoline; in addition, its customer profile showed decidedly older buyers of its products than the company deemed desirable. From a technical standpoint, because of the connotations of the Cities Service name the company and brand were frequently confused with either a utility or a municipal service. Both identity and design system needed to be changed. Lippincott & Margulies created an entirely new design system that enabled the service stations to be immediately noticed from the road, a new logo for the signs that was exceptionally powerful, and a new company and brand name.

For whatever reason, the name Citgo when first unveiled struck a wrong chord with a significant fraction of the driving public, who deeply resented it and raised an outcry for the new name to be rescinded. It was left in place, and within a year the rancor faded; in fact, a study done one year after introduction showed that Citgo stations had doubled the usual industry increase in annual retail sales. Over the years, the name Citgo became widely accepted, admired for its brevity and ease of pronunciation; in a word, it was memorable. In Kenmore Square in Boston a large Citgo sign dominated the skyline—and when, for economic reasons, the company announced plans to remove the sign, local residents mounted a campaign to protest its removal and have it designated a landmark, part of the city's architectural heritage. It was refurbished and relit, and continues to be celebrated by Bostonians. No one in their right mind would have predicted that Citgo's new identity would, over time, generate such a level of affection; at the inception, the challenge was just to get it accepted. After the initial resistance to anything new has been overcome, most names do become accepted, though most don't do so in as interesting a way as Citgo.

The Good, the Bad, and the Obvious

Aside from emotional considerations, there are other practical matters that go into the naming mix.

People have been conditioned to expect a name to be effective when it is first pronounced or first seen; more often than not, it is unreasonable to expect that this will be the case. Names by themselves are word symbols for what they identify; as symbols, then, they can't accomplish everything, though many people have unreasonable expectations of a new name in this regard. Few names have any inherent magic; it is how a name is used that makes it effective. Some, of course, do possess that wonderful quality of identifying the project and at the same time projecting a consumer benefit—among them Duracell, Eveready, Band-Aid, Sweet 'n Low, Taster's Choice, All-Bran. But there is nothing inherently wonderful about Cadillac, Tiffany, McDonald's, Colgate, Folger's, or GE; these are ordinary word symbols (for the most part, easy to pronounce) that have been made into classic, well-established, and highly valuable marketing tools by means of effective planned usage and associations.

In between these two groups is a third that consists of words that are new or adapted but have clearly been inspired by a natural association with the product itself. In this group I'll put Pampers, Xerox, Tropicana, Hartmarx, Jell-O, and

Gleem; each links itself to some logical feature, benefit, or heritage of the company or brand. Gleem is what the good toothpaste should make teeth do, for instance, and Hartmarx recalls the Hart, Schaffner & Marx brand of suits that for many years was a major factor in men's retailing. All these words convey messages appropriate to their products. One fortunate association-based name in this category is Goldome. After the Buffalo Bank for Savings had bought six or seven "downstate" marginal S&L's, it needed a new name for them; it just wouldn't do to emblazon "Buffalo Bank" on the corner of Fifty-ninth Street and Madison Avenue in Manhattan. At the same time it didn't want to dilute the market share it enjoyed in Buffalo. Its home in Buffalo had a gold dome, and we took this association for a name that "worked" for Buffalonians and that was perfectly acceptable (since it connoted money) in New York City.

A fourth group of words accomplish an even harder task: they come to stand for specific products despite their names. Chock Full O' Nuts is now understood to have nothing to do with nuts and everything to do with coffee—the company's market share is quite good. As a brand name for whiskey, Four Roses makes no sense on the surface, but it has been made to work. Aunt Jemima carried with it an unfortunate racial connotation that has for the most part been overcome. White Rock has little inherent imagery appropriate to a line of carbonated beverages. Smucker's, whose name connotes a derogatory slang word, overcame this problem by a clever advertising slogan: "With a name like Smucker's, it *has* to be good." Rambler as a brand name for an automobile didn't convey speed and excitement as many other automobile names did. Finally, the corporate name Philips Gloeilampenfabrieken, N. V. certainly would not have immediately suggested to most people one of the world's major electronic manufacturers.

Inherent weaknesses in names can be an insurmountable problem. The moniker of the California Packers Association was so mundane, unremarkable, and cumbersome that it was unable to serve a function beyond identifying a group of fruit and vegetable growers. It couldn't be used to market anything, so Lippincott & Margulies recommended the association adopt one of its brand names as its corporate one: Del Monte.

On the other side of the ledger, it only buttresses my case that some great brand names are not enough to ensure success. People's Express was an unusually apt identifier, but its promise was greater than the airline's ability to deliver on it. Zenith is about as great a dictionary word as can be owned—

but it has not been enough to sustain the brand when the appliances marketed under that name have been outgunned by the competition. Premier was a wonderful name for a new cigarette product, but the only thing likely to be remembered about RJR/Nabisco's infamous cigarette flop was the size of its failure. Clearly, a name doth not a product make.

Sorry, This Dictionary Is Empty

Of course it's not empty, but it seems to have been depleted of all the words you'd hoped to use for your new name; particularly annoying is to discover that the really good ones have previously been snapped up by your direct competition.

As a consequence of most appropriate dictionary words having been "taken," or legally registered by other companies, many name candidates today are fabricated words, such as Meritor, Unisys, Ameritrust, Tylenol, Kotex, 7-Up. Unlike the famous names cited in the paragraphs just above, these words take some effort to be made to work for a corporation, and quite a bit of additional expense (and time) to make them meaningful to the corporation's target audiences. But it can be done. Once, a client who was facing a rigid deadline for picking a new corporate name had its legal department limit its choices to Xylex or Celeron, neither of which the client liked. With time running out, the client chose Celeron; over time that peculiar-sounding, completely fabricated name came to serve its purposes quite effectively.

It's important to act quickly, for many names—even made-up ones—are disappearing from the list of possible combinations. To our astonishment, the Primerica name had actually been taken a couple of months before our client tried to register it. The only solution for the client was to buy the rights to the name from the small Maryland financial services company that had just registered the name but had not yet used it.

In the case of another company—fortunately, not a client of ours—the chairman announced his new name with great hoopla at a well-attended stockholders meeting, only to learn shortly thereafter that another company owned clear title to the name. The unlucky learnee was USX. Then the former United States Steel Company, having so publicly introduced its new name, had no choice but to negotiate with the legitimate owners of the USX name, a small telephone service company in California. While I don't know the details of the settlement, there is no doubt that the small phone company became a lot richer, and quite a bit faster than it might have thought possible.

When international rather than domestic protection is required, the legal obstacles become even higher. Worldwide registration takes a great deal of time; few clients want to wait to complete the whole procedure before actively using the name. Also, the chances of any word being equally legally available in all of the major market countries are indeed slim. One company tried waiting for the entire clearance procedure, and investigated the availability of nineteen name candidates in twenty-five countries; only two survived.

Translating name candidates into foreign languages is an increasingly important, though danger-filled, part of the naming game. Years ago, it was the Romance languages that gave primary cause for concern, but today there are even more languages to consider. The Cadillac Division of GM asked us to determine whether Allante was offensive when used in a foreign language; we were asked to check twenty foreign tongues, with an emphasis on various Arabic ones, because of the popularity of Cadillacs among Arabs.

The Pet Milk Company ran into difficulty when trying to change its name to Pet, Inc.—pet in French means flatulence, and one of the company's major markets was in the French-speaking area of Canada. In the days before international marketing assumed its current importance, that would not have mattered; today it is of increasing importance to every company. More recently, PetroCanada approached us to develop a new brand name, since ownership of the company was changing from private to government hands. One of our most logical candidate names was Petron, but it couldn't even make the short list, since it could be translated as a "pet" derivative and was therefore not even a remotely acceptable possibility for a product to be sold exclusively in Canada. Similarly, a conservative, old-line, prestigious bank was in the final stages of choosing a new name when we discovered that one of the prime name candidates, when translated into colloquial Spanish, meant a large male organ.

The greater the globalization plans, the more difficulties arise. Chevrolet was unable to sell its hot new car in Mexico because there Nova was pronounced as "*no va,*" which means "does not go." A similar problem perplexed us when we recommended Paccar as the new name for a company formerly known as Pacific Car and Foundry, manufacturers of heavy-duty equipment and those huge trucks known under the brand names of Peterbilt and Kenworth. Paccar was appropriate because of its link with the company's past. It had only one drawback, which in the 1970s was not seen as crucial: when translated into Russian and spelled the Russian way as Paczar, it meant the Czar of Rats or Rat of the Czars. The new name was adopted; today, it might not have been so readily approved, since it could compromise a potentially large Russian market for the company.

Names as Unifiers

In the mid-1980s, Houston Natural Gas and Internorth merged to form the largest natural gas company in the United States. It was a logical and not a hostile merger, greatly to the benefit of both companies. The new company faced some interesting problems and challenges. It was automatically going to become a leader in the energy industry, so its name had to be appropriate to that position. It also had to convey the idea that the company was indeed in the energy business, rather than only in the somewhat narrower, recently deregulated natural gas business. For a host of reasons, most of them having to do with the internal politics of both companies, the company also needed a name that would be a unifier for the two very different corporate cultures of the partners.

There was a great deal of bitterness and controversy within the corporate cultures—over who had bought whom. This anger spread all the way up to the boardroom. At one particular afternoon board meeting to which we had been invited, we were shocked to discover that the chairman who had invited us was no longer chairman; in fact, he wasn't even in the room, since he'd resigned just before the lunch break. And he was the second chairman to go since the merger had been announced.

To make a successful future, the company desperately needed a unified work force. We recommended that the new company break entirely with the past and adopt a new name that would help erase the memories of old rivalries and what we call the "we-they" syndrome—"We're from Internorth, they're HNG." We needed a name that would not perpetuate the difficulties. From a list of potential candidates we had compiled, CEO Ken Lay selected the name Enteron. There was a dictionary definition for it, but only a medical one, rather obscure, for a word relating to the alimentary canal; we advised our client of this circumstance, but also said that we thought it was not (you should pardon the expression) an impediment to its being adopted. Actually, we first advised Lay to hold back on the recommendation until we'd had enough time to develop a logo expression. He decided to go right ahead without that because, he told us in his best West Texas drawl, "you guys can make a dawg look good."

Not that time. No amount of cosmetic uplift could make this "dawg" do more than say "bow-wow." A strong chorus of complaints moved us to recommend a slightly modified new name, Enron, which Ken Lay and the board approved. As we all had hoped, it became a standard to which both former cadres of executives could rally, and helped ease the internal political problems that invariably arise from the merger of two large companies.

Just to show you that a new name is not always the answer, even to a unification problem, let me briefly tell the story of two neighboring gas and electric companies in Ohio that decided to merge to take advantage of the operational economies that would ensue. We discovered there was so much loyalty to Toledo Gas & Electric and its counterpart that a single name unifying them would cause problems. So we fashioned Centerior Energy for the name of the holding company (a name that would be listed on the stock exchanges and would have its primary appeal to financial analysts), and left the two companies, names untouched, as subsidiaries.

In the instance of the First National Bank of Jackson, Mississippi, there was a need for a name change to expand a territorial base; the solution also became a unifier. The bank had bought a lot of small entities scattered throughout Mississippi and neighboring areas, and since there were "First National Banks" in many of those cities, knew its own name would not suffice to use on acquisitions. We evolved the name Trustmark National Bank, and applied it to all the branches through a compelling design system. So far, so standard. When an independent research company looked into the new bank's image a year later, it found some surprising (and to us, gratifying) results. Consumers thought that the newly named bank had grown in size, although it actually hadn't, since the number of branches remained the same; they also believed that service had improved. Indeed, service had improved because of the "unifier" principle. Employees felt better about belonging to and representing this newly named entity whose single name was now spread over many small towns; employees also benefited from the prestige that came to an organization perceived as larger than it actually was.

Over the Hill?

While our responsibility in naming generally ceases after we receive legal reassurance that a name is viable, we frequently advise our clients on how best to make sure that title to their new name does not become compromised over time. In some cases, loss is not a matter of the name growing old or connoting a bygone era, but rather of sending a name over the hill by inadequate image management practices. In recent memory the original manufacturers have lost the names Nylon, Escalator, Aspirin, and Cellophane. These words were originally invented to describe new inventions and new categories of products, but now they have become generics. They can be used by other manufacturers to describe their own versions of the particular product.

Some brand-name words that could have suffered the same fate but did not include Band-Aid, Jell-O, Kleenex, and Xerox. These words have been kept alive as proprietary names because of diligent legal supervision and, above all, recognition by management of the importance and significance of maintaining the exclusive rights to the name. The obvious value of these several product and services names provides more evidence of how important it is for a company to consider its actions beyond the introduction of a name and image; maintenance of the brand's prominence and integrity is a key factor, and requires continuing monetary commitment by management.

DuPont even learned from past experience. After the company had lost the exclusive rights to two of its key names, Nylon and Cellophane, it wisely determined not to lose another name in the same way. Having invented a new stretch fiber, it developed a generic name for it, Spandex, and gave that name to the industry as a whole. DuPont allowed anyone to produce a stretch fiber to its specifications and call it Spandex. But it trademarked and promoted its own particular brand of Spandex called Lycra, and put its marketing efforts into that rather than into the generic. As new inventions enter the marketplace, inventions that will by virtue of their novelty and uniqueness require new words to describe and name them, strategic planning on how to keep a name from going over the hill—either because of age, by becoming a generic, or by being somehow usurped by the competition—will become increasingly imperative.

Nomenclature Architecture: Overlooked, Underdeveloped

Much has been written about the role of graphics and graphic design systems in the shaping of images. But less understood are the consequences of linking one word with others, or any of the other facets of nomenclature architecture, such as the need to articulate policies and procedures for name development and word usage in the development of communications strategies. Most of the time, names seem to be given to products and services without regard to how they relate to the company as a whole, or to other products and services of the same company. Someone has a brilliant idea, that name is researched and adopted—often after considerable expense—but the work is done in a vacuum, as if it does not relate to anything else the company is doing. The consequences of ill-conceived nomenclature architecture are not well understood, though very often companies suffer from them.

Companies that are successful with their nomenclature architecture usually have a well-established pattern for naming their products or services. Procter & Gamble's policy is to completely separate each of its brands from the others by means of name. Gleem has nothing to do with Tide, Pampers nothing to do with Scope, Ivory nothing to do with Pringles. And none of these is in any way communicatively linked to the Procter & Gamble corporate identity. This strategy of total separation allows each brand to develop by itself, and to the maximum extent possible without impinging on any of the others. Such isolation and concentration often result in extremely strong individual brand performances; it is hoped that each brand will become dominant in its own field, but if it does, that will not necessarily positively affect any of the company's other brands. The difficulty with complete separation of brands is that funding the strategy of fully supporting each individual brand is extremely costly, requiring a continued high level of spending to sustain them all. Consequently, the separation approach is an option only for companies that can afford it. The very expense of introducing and building a new brand with a completely new name is a definite inhibiting factor in the marketing of new products when the separation approach is followed.

Johnson & Johnson has followed a nomenclature approach based on product function. For instance, its strategy forbids the use of the Johnson & Johnson brand name on any product that is to be ingested. Hence, while one can buy Johnson & Johnson's Band-Aids, Baby Shampoo, or Dental Floss, one cannot buy Johnson & Johnson acetaminophen tablets—the latter are known as Tylenol. When a new product is to be introduced under the system, it is done through classic line extension; thus, CoTylenol, a name associated with the older brand yet still separated from the Johnson & Johnson corporate name by the system of nomenclature architecture. While one may agree or disagree with the line of reasoning that has produced this system, it is admirable that at least a system has been thought out and used as a guide to policy.

Sony has a system, too; it has a double tier of names. Over the years Sony has built its image not only as a corporation but also as a quality brand name that is applied across a broad range of products: televisions, portable radios, hi-fi equipment. Sony brand also encompasses a variety of prices, though Sony's products are almost always expensive and the best-designed in their categories.

The second tier has to do with the creation of specific new brand names for new products. The example here is the Walkman. When Sony introduced this new personal portable stereo, it created a new brand name—Walkman—that projected image characteristics based on its use, and then insisted that it be linked through name to the master brand: the product is known as the Sony Walkman. The name is a triumph of brand leverage. It positions the product appropriately; it lets the Walkman's own brand image feed off the already-existing good image of the master brand. And since the product is a good one, the Walkman (and its name) reinforce Sony's reputation for innovation and excellence in consumer electronics. By establishing both tiers of names, Sony also makes easier any line extensions, such as the Sony Video Walkman, a product that was already in the pipeline when the first audio Walkman was about to be introduced.

Let me note in passing that not all consumer electronics companies have similar nomenclature architecture systems. That of Philips is considerably less unified, veering as it does toward the Procter & Gamble model in which each product stands alone; the problem here, as I've pointed out just above, is that it takes a lot of money to sustain many individual brands. That's why Sony's marketing efforts, less costly because of the recognition of its master brand name, have a head start on the competition. A secondary problem, common to many electronics manufacturers, is a reliance on what is called "'alphanumeric" naming. These manufacturers expect consumers to know the difference between a product labeled as, say, A320, and one called by the equally unedifying moniker of G561. (Some automobile companies have fallen in love with alphanumerics, too, but this is less dangerous to them since they have far fewer models and numbers to worry about.) Leaving the decipherment of a name up to the consumers simply gives them one more reason to scratch their heads in disgust and buy the correctly named brand they already recognize.

Stretching

Because the cost of introducing and marketing new brands is so high, many companies have tried to "stretch" already-existing brands to accommodate new entries under old wings; the technical name for this practice is line extension. The point is that consumers already recognize one name, and, since their attention spans are limited, they'll be more apt to choose a product that is under the umbrella of a familiar name than one that is entirely new. "If you get brand-stretching right," says one of the marketers of Hellmann's mayonnaise and Mazola oil in an article in *The Economist* (May 5, 1990), "you can travel further for less money." But, he warns, "if you get it wrong, you risk weakening the core values of the original product."

One can't go too far wrong, though, because well-established brands are fairly resistant to damage, even by badly handled new products that bear their old names. Even the disaster of New Coke didn't truly damage Coca-Cola. So the controlling factor seems to be costs. A London marketing consultant's study cited in the same article suggests that advertising and promotion costs were 36 percent lower for stretched brands introduced in the last five years than for wholly new brands; moreover, the consultants found, stretched brands have a higher survival rate than entirely new ones. Maybe, if they are managed properly, the stretched brands thus introduced will endure for as long as some of the really good ones. Nielsen did a survey and found that of the top twenty-two brands of the year 1925, nineteen of them still led their product categories sixty years later.

Calling All Brands

Without careful planning, many opportunities can be lost—chances to establish a world-class brand, to make product introductions simpler, to save on costs through an already-established system that can leverage a new product from the equity built up by earlier ones.

AT&T learned this lesson the hard way.

When the company was broken apart in 1983, there were many issues facing it, and during its first years as a non-monopoly the company was wholly fixed on problems of how to retain and regain business. Marketing and in particular the building of brands was lost in the shuffle. During these years, companies introduced many new products and services, most of them doing so without having in place any particular policy to guide marketing managers in how best to name and position these new products and services. Among the new ones of this era: Calling All America, USA Direct Service, Teleplan, and Trimline. When introduced, they had two things in common: first, all were provided by the best telecommunications company in the world; second, there was no mention of that company and its splendid name (AT&T) in the brand name of the new product or service.

Further, the company branded some of its other products with either unregistered names or generics: Memory Telephone 530; Typewriter Model 6710; collect call; interstate directory assistance. All of these brandings ignored the extraordinary value of the AT&T identity.

Only after a half-dozen years was a careful set of policies developed to insure that the company's equity in its own name would be transferred to its products and services, and that new services would be subject to naming through a coherent no-menclature architecture plan. Under this new policy, a conscious effort is being made to build AT&T as a master brand. This idea follows the notion, often stated in this book, that a company's name and its image are among a corporation's most valuable assets, and that precise policies are required both to leverage from and protect this equity.

Under this new policy, AT&T as a name precedes the formal name of all products and services. Reach Out America became AT&T Reach Out America; USA Direct became AT&T Direct. When a new credit card was introduced, it was named AT&T Universal.

Now the difference may not seem all that significant at the moment, but it is an attempt to build something, and to position each service as part of a larger dedicated telecommunications service. The nomenclature architecture facilitates cross-marketing opportunities, and allows for realizing cost savings in promotion and advertising budgets. Think, in this regard, of what McDonald's has achieved by having its considerable line extensions all introduced through the "Mc" connection—Egg

McMuffin, Chicken McNuggets. The success of the new product reinforces the master brand.

I do not necessarily advocate one system of nomenclature architecture over another. Totally independent standing identities of the Cadillac-Oldsmobile-Chevrolet sort can be as effective in their way as the linked names of the Mercedes line. The imperative is for nomenclature architecture systems that are thoughtfully developed and crafted, systems that can help a company now and especially in the introduction of future products and services.

PART 4

Recommendations and Admonitions

CHAPTER 13

Beyond Mass Marketing

Image making is an extremely valuable way of looking at many of the problems businesses will increasingly face in the future. How to prepare for the future, how to cope with budgetary changes, how to evaluate a company's assets, how to deal with new technology, how to defend against a hostile takeover—these "hot topics" are under intense discussion and scrutiny in boardrooms. By approaching them from a communications standpoint, we can put them in a new light.

As many business leaders have recognized, the era of simple mass marketing is coming to an end. In order to sell deodorants, it used to be enough just to announce the existence of your product, give people a reason to use it, and have proper distribution channels set up to handle the flow. No longer. Now—and, increasingly, in the future—the marketer of deodorant needs more than one product with one smell. It is preferable to have an entire line of smells differentiated into sticks, balls, aerosols, and dry sprays, and compartmentalized into products designed to appeal to men-in-the-evening, to athletes, to women-on-the-go, and even one that will be just right for couples who share the same bathroom. In short, future products will have to be marketed to segmented groups, and in a more responsive way.

Moving toward the Luxury Market

At the turn of the twentieth century, Henry Ford saw automobiles as the quintessential mass-market product. If the production techniques could be perfected so that the price of a car could be brought into the realm of possible purchase for every American household, Ford reasoned, his fortune would be made. And it was. And major carmakers followed Ford's route for many years—including the successful Japanese carmakers, and especially the Nissan Company. From the 1950s to the 1980s, Nissan made its money from selling low-priced models that competed successfully with Detroit's for consumer dollars. As the 1980s drew to a close, though, in its research on future sales Nissan learned from its own projections that it would not be able to make a good enough profit margin or sustain its remarkable record of growth any longer only on the basis of selling low-priced cars. Also, as its satisfied customers climbed the ladder of economic success and wanted to let the world know how they had prospered, many of them, in truly American fashion, looked to buy a more expensive car, and switched to a competing company—Mercedes, BMW, Jaguar—because no luxury Nissan product was available. Thus, previous satisfied customers were lost to the company, resulting in a waste of goodwill. The company wisely decided that the future lay in going beyond its mass-marketed cars by adding something new: luxury cars designed to sell for considerably more money than Nissan's regular line. This thinking led to the Infiniti, whose story I detailed earlier.

But I want to make another point about Nissan's reasoning: it was not that the market needed another luxury car to compete with Mercedes-Benz, BMW, Jaguar, and Cadillac—but, rather, that Nissan required a move into the luxury market in order to continue to grow as a company. Then the question became how best to enter that new market niche and project a different image than its well-entrenched competitors.

To establish the Infiniti line was a gamble for Nissan, of course, but I'm sure that in the long term it will pay off. And that is the real point of this story, the emphasis on the long term. Nissan understood that image making on this scale requires a long time partly because Japanese companies have a greater commitment to long-term strategies than American companies do. Nissan reasons that the hundreds of millions it is spending to introduce Infiniti will still be paying off in terms of corporate profits many years into the future, because its identity practices are firmly establishing a product whose profit margins are great enough to sustain Nissan into the next century.

As you will have realized, other car manufacturers must have been making their own, similar projections about the future profits to be reaped or not reaped from mass-marketed automobiles, and in the last quarter of 1989 began to do what Nissan had done—but with a difference. Ford Motor Company purchased a stake in Jaguar, for a very high price—a price so high, in fact, that commentators almost universally pointed out that Jaguar had to have been purchased not for the profits it might bring on Jaguar sales, but for the cachet Jaguar's luxurious image might add to the mass-marketed lines of Ford. An American car manufacturer, more impatient and less fond of risk than the Japanese competition, wanted a top-of-the-line product, and wanted it now, and so has paid a huge premium to purchase 1) the advantage of collapsing development time schedules, and 2) the security of knowing precisely the status and prestige that its new luxury line—Jaguar—possesses. The only thing Ford risked was the premium over market price that the company paid for the Jaguar company. In fact the purchase was an avoidance of the risk of *not* having an entry into the prestige sector of the market. Nissan went one way, Ford another. Regardless of which company may later be awarded the palm of having made the better guess, it is clear that image was the essence of both routes to wooing the luxury-car-buying public.

The combination of carmakers goes on; begun earlier in the 1980s with alliances between American and Japanese makers, it continues in the 1990s with marriages between and among European, American, and Japanese manufacturers to produce automobiles increasingly aimed at precisely defined segments of the car-buying public. While such alliances seem to be good ideas from a strictly financial point of view, I am doubtful about their success unless they pay more attention to positioning brand identities with precision and clarity, so that car buyers have an understanding of what each brand represents. While the appeal of Infiniti and Jaguar is precisely aimed, that of the newer amalgamations is murkier, and thus may fall short of expectations.

New Pastures

If products and corporate amalgamations such as those noted above do fail, it may well be because they have not understood the new demographics. In the post-mass-marketing era there will be many changes in the target populations that consume the products of American (and global) industry. Customers are already more sophisticated and knowledgeable than they were in former years. One can now sell quiche in Kansas, expensive Bordeaux wines in South Dakota. But to do so, one needs to send out advertising and other messages that are specific to the market niche. Once upon a time, Kraft Foods could make a product, call it Sliced American Cheese, and obtain for that product a huge percentage of the cheese-eating market. Now, in a good supermarket, fifty cheeses compete for space and customers. Kraft's Sliced American Cheese being only one of those products. To sell cheese in the future, the marketing of it and its image must be more adroit and specific to the customer. How so?

1) To locate and exploit new markets for older products, manufacturers may have to *look to entirely new pastures*—for example, overseas, or in cultures different than that of the United States. In image terms, this will mean producing or cultivating image characteristics that travel well, and that make sense in a variety of cultures. Remember that in Italy, pasta was such a staple food item that it might as well have been considered Italy's sliced white bread—but when the image of pasta was reshaped for consumption in the United States, relating it to the American taste for innovative, low-fat foods, its growth in acceptance here became remarkable. Not so, a staple of the Scandinavian and Eastern European diet—the herring. Despite the food fish's popularity with millions of people in those countries, the herring's image was never reshaped to achieve wide acceptance in North America, and has never sold as well here.

2) To sell new products, images must be designed that stress the *appeal to highly specific or restrictive market niches*. For instance, a decision might be made not to sell a new blue cheese to any fast-food chains, but to reserve it only for the "limited doors" of upscale gourmet boutiques.

3) If mass marketing of an established product is to continue as before, *care must be taken so that the image of the product does not slip from the category of a premium product to that of a commodity*—for example, as has happened (to its detriment) with Swiss cheese. Now, to paraphrase a slogan, "You don't have to be Swiss to make Swiss cheese."

Though many marketers have understood the notion of marketing to segmented groups, they have not always assessed the consequence of the demise of mass marketing. Also eclipsed in that development were antiquated corporate and brand identity management practices.

They gave up the ghost because they represented outmoded attitudes about media use. For several decades, the prime medium for building an image for a product was network television advertising. The classic example is Revlon. In the early 1950s Revlon was a rather small company. But through its sponsorship of the TV program "The $64,000 Question," Revlon took on the glamour associated with the television medium itself, and grew to become a major force in the cosmetics industry. In that simpler era, an advertiser didn't have to concern itself with worries about what media to use to tell its story: network television was the locus of all the action. Products advertised on prime-time television acquired an aura of success as well as glamour.

Medium as Message

Only Marshall McLuhan realized quite early on that, as he wrote, "the medium is the message." All media transmit their own image-affecting messages: products advertised in tabloids that are sold in supermarkets are often thought of as tawdry; those sold via matchbook covers are considered cheap. The future of image shaping will present more complex problems than heretofore, because the media have multiplied and are fragmenting, offering a broad range of options and mixes. Let's take a moment and review some of the newer media.

1) *Narrowcasting.* Major network television has been labeled broadcasting, so cable television positioned itself on the opposite end of the spectrum by appropriating for itself the label narrowcasting—that is, it became a collection of stations each of which appealed to a narrow segment of the audience. An important moment was reached in the summer of 1989: during one week for which ratings were taken, the combined "share" of the total viewing audience that watched programs on the three major networks dropped below 50 percent for the first time; that is, as many people were watching local stations or cable channels or using their television sets through their VCRs as were watching the networks. In the future, can a company interested in appealing to the general public afford to write off 50 percent of the viewing audience by ignoring the cable viewers?

2) *Home video rental.* Many makers of videos for rental are now including advertisements within the cassettes. Their attraction for the advertiser is a captive audience with known demographics. Should the medium remain relatively unregulated, it will offer potential for targeted advertising of a sort hitherto unrealized by the television networks. For instance, a movie that appeals to a teenage audience might attract advertisers for beer, soft drinks, pimple cream, and safe-sex products. A sophisticated romantic comedy? Just right for the marketers of liquor, expensive automobiles, and exotic travel. Serious drama—equated with a well-educated audience—could be particularly attractive to the Book-of-the-Month Club and to classical record labels.

3) *Supermarket messages.* New shopping carts displaying video messages, and loudspeakers emitting audio messages are being placed in the most advanced supermarkets being built today. These provide the shopper with input right at the point of sale—information about the products on the shelves as well as messages urging the purchase of particular items.

4) *Wallboards.* With Whittle Communications as the pioneer, sponsored wallboards are being placed in hospitals and in the offices of doctors and dentists, and in high schools, pet shops, exercise parlors, and other places where distinct and definable groups of potential customers often congregate.

5) *Computers.* Messages about products and services now go out over networks of computers. Also, some advertisers are now shaping messages specifically for computer users; in a recent computer-user magazine, Ford advertised a low-cost disk that contains "terrific graphics . . . Exciting games to put you in the driver's seat," information on the various models of cars made by Ford, and "a friendly spreadsheet [to help you] figure out the cost of your dream car."

The multiplicity of the new media tends to fragment an image. The glamour you might once have obtained from sending your message out on network television is distorted and possibly debased if the same message plays on a radio in a drugstore. Consequently, the need is to develop image-making tools that can preserve a particular image regardless of the medium in which the message will be communicated.

CDVCRPCISDNHDTVWORM

No, these letters don't add up to gobbledygook; they are specific identities for mediums that will be used to deliver image-shaping messages in the future. They are tomorrow's printed page, the channels available to corporations for communicating with their audiences. During all of recorded history prior to the invention of the Gutenberg press, the conveying of messages was dependent upon each book being slowly inked by a single individual; to obtain more than one copy, communicators had need of as many monks to ink the copies. The Gutenberg press changed the character of communications by making it possible to manufacture and disseminate multiple copies of a message. The compact disc (CD), the videocassette recorder (VCR), the personal computer (PC) the Integrated Services Digital Network (ISDN), high-definition television (HDTV), and a new version of a Write Once, Read Many (WORM) disc will provide copywriters, art directors, commercial producers, and designers with tomorrow's equivalent of the printed page, and their impact may be every bit as revolutionary as was the introduction of the printing press.

Consider the difference in communications between a magazine advertisement and a message that comes at you on a PC screen. The magazine is held in your hand, where you can feel the smooth, glossy pages and allow your field of vision to be completely filled by a two-page advertising spread; you have a tactile and quite personal relationship with the medium, and that helps shape your views of the images and selling messages aimed at you through the magazine's pages. In contrast, the PC screen is cold, sits at a distance from you, constitutes an environment that may be considerably cluttered with many competing messages, and may well be non–user-friendly. Most probably, you relate to the PC exclusively as an information provider, and may not yet be comfortable with the idea of receiving image-shaping messages from it. But that will happen in the not-too-distant future. As a consequence of technological innovations and improvements, there will be increasing crossovers enabling television screens to function also as computer screens, and marrying compact disc players with television and even with telephone switching systems to provide information and entertainment interchangeably.

Right now, some football fans contend that high resolution cameras, instant replay, and the whole host of other technological tricks used in the network broadcasts of big-time football games offer the fan at home a better overall view of the action than it is possible to obtain from being present at the stadium where that game is being played. On the other hand, opera lovers complain that televised broadcasts of opera, on small television screens and with the moderate sound quality available from most television sets, are a poor substitute for attending a live performance. But suppose those opera lovers could see a well-televised performance on a high-definition screen that was the size of their living-room wall, and simultaneously hear its sound on the equivalent of a CD player—well, then the home viewer of opera might feel more satisfied with experiencing the opera at home and not paying the increasingly high price of good seats at a live performance. When that technological eventuality occurs, you can bet that companies seeking to shape images will line up and pay well for the chance to send messages to those viewer-listener-experiencers.

Or consider the possibilities of interactive television. In an experimental project in certain cities in Canada today, when a selected group of viewers watch and listen to the evening news, they do so with remote control buttons in hand. After the initial section of the news, say, the first eight minutes, in which the major stories are headlined, viewers are then presented with a choice. They can go on with the regular presentation, or choose to watch in-depth presentations on one of the top stories, or get more sports and less weather, etc., by pressing buttons on their remote control devices. The result is programs that are shaped in some active way by viewers. Advertisers are willing to pay premium rates to transmit their messages to those people who, for example, make active choices to obtain more business news, or more medicine and health stories, by the click of a remote control button.

Similarly, when one calls Fidelity Investments' 800 number from a touchtone phone, the caller is requested to touch particular phone buttons that will branch his or her requests; after a few electronic queries from the telephone to the caller, and the caller's push-buttoned answers, the caller reaches the precise information-provider who can best respond to his or her requests, whether they be for information about a fund, quotes on current prices, the way to apply for a retirement plan, and so on. Note that when the caller and ultimate information-provider are connected, the conversation proceeds on a very specific level, with great density possible in the transaction. Here, too, interactive choices are being made and responded to in important new ways.

A recent report in *The Economist* (March 17, 1990) announced that "multimedia" products will be arriving in consumer electronics stores in late 1991. These products will combine the PC, the CD, and television to make "a computer that can both show and tell." One already announced program

> lets the viewer stroll through the Smithsonian's galleries. By pointing to, say, a spaceship, the viewer can call up film of the launch or biographies of the astro-

nauts. A multimedia edition of Grolier's Encyclopedia works in a similar way. Also on offer will be a multimedia biography of Frank Sinatra, a video jukebox, and a disc called "The Sexual Universe" (be gentle with that joystick).

An even further-out technology, to come on-line in a few years, is currently being developed by more than forty Japanese companies. It is called the MO, the magnetic-optical disk, described in another *Economist* news article (May 12, 1990) as "the CD's grandchild—a handy-sized laser disc that can store millions of words, pictures, musical notes or computer data." The present generation of MOs is very expensive, and has a drawback—the MOs are known as WORMs, for Write Once, Read Many times, since they are difficult to erase and reuse, but once material is put on them it cannot easily be removed or replaced. One WORM can store the equivalent of a small library of books—say, as many as a college student could use in four years of courses. The new MOs will also be more easily erased and reused, and are even larger; a single MO may soon hold the texts of one thousand paperback books. The new computer being designed by Steve Jobs, who did so much for the Apple and Macintosh lines, will contain an MO drive—an indication of the fabulous possibilities in this new technology.

Messages aimed at people who have made interactive choices for their PCs or television screens (or in their telephone queries) ought to be exceedingly well thought-out and focused, or else a big and expensive opportunity will be wasted. What I'm trying to demonstrate is that in these new high-technology media there is great promise for future image-shaping activities; but since the environments in which messages will be received will drastically change, image shapers will have to adjust their tactics to take advantage of the new media—or risk being left behind.

Let me emphasize that this is not a technical problem. There are fewer and fewer computer illiterates active in the business world, and technological expertise will be ubiquitous enough so that no company need be lacking in technical understanding of the new media. No, the danger lies in what T. S. Eliot used as a definition for hell, the place where "nothing connects to nothing." To avoid that lack of connection, image shapers will be forced to comprehend and make creative use of the possibilities of the new media. The elements of identity tools themselves will remain constant—words, symbols, sounds, colors—but new ones will be added and become more important, and the emphasis on older ones will shift. There will be more variables affecting the delivery of a message.

A characteristic of the new technologies is the ability to reproduce the smallest details; these details result in a greater sense of realism. But the opportunity exists within the new technologies to do something further with this sense of realism. Historically, only those things that could be photographed or illustrated could be projected, but in the future the details that can be so realistically limned can be imagined ones rather than those that actually exist. In the future, whole scenarios can be imagined and brought to life (in a most convincing manner) through details so real that viewers may well be persuaded to adopt whatever point of view is being suggested. Such a powerful level of graphic presentation will obviously provide image shapers with new palettes wonderful to contemplate.

If that is too far-out for you, let's bring into focus something closer to the present. How about your corporate logo? Most logos today are designed for print. They are frozen images that look good on corporate letterheads and in magazine advertisements, but which have hardly begun to take advantage of the electronic media. Only a very few logos have to do with the single most effective element of television—motion. One that I admire is the Burlington Mills logo, a crosshatch pattern that interweaves into a fabric as you watch: a true moving image logo. AT&Ts new swirling and evolving world-ball is another one that takes advantage of the medium. I suppose they all owe a debt to MGM's roaring lion, an early animated logo, and to the Westinghouse logo with its lighting bulbs, a corporate symbol that was imaginatively translated from a print to a moving image logo. And what of sound, an almost neglected element in today's logos? The transmission repair expert AAMCO, with its "Double A—beep, beep—Em Cee Oh" signature, is one of the few that has a well-used audio element. In the future, the dominant elements in image shaping through these new media may not be the ones emphasized in print; my guess is that sound and motion will be incorporated in the most effective of tomorrow's messages. Today, most of us in the business of shaping images are print-oriented. Tomorrow, that just won't do.

Many identities originally designed exclusively for print will have to be modified to benefit from the new dimensions afforded by the new media. The visual personality of a corporation should be extended to embrace all media, so that when an image pops up on a screen—whether a PC or a wall-sized TV—it will be immediately recognized. For instance, if there is only a small logo visible at the end of a message, it may be lost in the clutter of information and be of no use. Consider the problem of how to make your message prominent in, say, something like electronic mail. The portions of a message that appear on a screen at any point in time won't necessarily display the company's letterhead, as would happen if the reader held a direct-mail letter in his or her hand. To make sure that the message gets across, a way will have to be devised so that any portion of the direct-electronic-mail letter on someone's home PC screen will project a distinctive look that is proprietary to the transmitting company. Graphics will be redefined to include the total style of the communication and the media in which the messages will appear. The task is eminently feasible, since the tools are extant; the problem remains the leap of imagination necessary to incorporating knowledge of the new media and their possibilities into our thinking.

CHAPTER 14

Bottom Lines and Book Values

A corporation's bottom line is, of course, its most important number; this has always been so, and will continue to be so. And bottom lines are expressed in nice, solid, uncompromisingly precise numbers. What I shall argue in this chapter is that in the future we must come to an understanding that the corporation's bottom line is not *only* made up of numbers—that there are components expressed in other ways that are neither less important nor less real than numbers.

Quantification—The False God?

Sam Wanamaker, founder of the Philadelphia department store chain, was famous for saying, "Half the money I spend on advertising is wasted, but I don't know which half." The phrase actually dates back to the eighteenth century and is usually cited in praise of advertising. It suggests that advertising's effect is difficult to measure—and, therefore, that the advertising agencies ought not to be pressed, because to do so would be to disturb the magic. Those who cite the saying contend that Sam was fortunate to have had advertising that worked, and should not have looked for a way to cut the cost in half.

I interpret the saying differently. To me, it is evidence of the good American corporate executive's continual frustration at not being able to measure in numerical terms the services he or she is paying for. In fact, our corporate executives are continually searching for ways to measure aspects of their business. Numbers are the lowest common denominator in the problem of measurement; moreover, American executives are trained in numbers—and so executives tend to look for numbers, even in places and processes that are not particularly susceptible to numerical measurement. Let me state categorically that the great American corporate game of "show me the numbers" is grossly overvalued and misused when applied to measuring image. Now I'm not saying image can't be measured or that market research does not have an important role to play. Market research is often quite valuable, and image can be measured—I'll discuss how, just below—but traditional financial equations are not the proper yardstick.

Management often looks for quantification in difficult areas, to help it make a decision. Market research, for instance, is a staple of American business. Because the stakes involved in marketing new products are so high, executives try to find ways of backing up their positions with numbers—any numbers. In case things go wrong, executives like quantifiable evidence to support the decisions they made, so they can't be blamed and fired for making the wrong decision. An example is the introduction of New Coke. Before marketing this new product, the company did extensive blind taste-testings. Groups were asked to taste several different products, and in every tasting New Coke proved more popular; people said they preferred its sweeter taste to the other soft drinks proffered.

I'd been trying to tell cola companies for years that their main product was not the taste of the sweet syrup with fizz, but image; just as the most important entity sold by GM in its luxury car line was not four wheels and an engine—and a great product—but the Cadillac image. My point was that the precise taste of the product didn't matter, and in the case of the introduction of New Coke, this point was spectacularly proved, as consumer backlash soon forced the Coca-Cola Company to drastically alter its plans and to reintroduce as its flagship product what it now styled Classic Coke. In this instance, market research proved to be without value. Similarly, market research done by motion picture companies told executives that they should not bring out either *Star Wars* or *E.T.*, two films that have garnered among the half-dozen largest gross incomes in the history of Hollywood.

Actually, there are statistical measures of the efficacy of advertising and promotions—coupons returned through retail stores, Nielsen ratings, the rise or fall of a product's market share, the percentage of hits to throwaways in a direct-mail campaign. These measurements are comforting. You also buy advertising in terms of how many households you are reaching or in terms of so many dollars for a certain group of potential consumers—say, women under forty, or males with high incomes. By spending four times as much as a competitor does on advertising for a similar product, you can probably increase sales quite a bit; the question then becomes, has the number of increased sales justified the additional advertising dollars spent? But in terms of image, the same numerical ratios are not applicable. You can, indeed, buy an image, but you can't buy it by the pound. If someone has paid X dollars to a consulting firm for an image, is that image half as good as that of a company that paid 2X for similar services?

Elements of corporate and brand images are measurable—they just aren't always quantifiable in precise or even relevant numbers. How can we measure images? Through questionnaires and well-developed interviewing techniques. We do so all the time, in trying to find out the attitudes a company's various audiences have toward it. Trying to understand and measure the efficacy of a company's image is our meat and potatoes, and we want to have as much information on it as we can obtain for a client's benefit. But the techniques we use have to do with quality, not quantity. For instance, we question members of the financial community as to how they view a certain company. But if we personally interview three financial analysts, do we get less of a usable response than if we ask ten simply to fill out a questionnaire? The utility of the information depends, of course, on the insight and quality of the analysts involved, and the time they spend on giving us honest answers—which is another way of saying that numbers aren't the whole story in this regard, either. We also measure image in terms of awareness levels for a product, or in terms of how well or badly a new product is accepted in the marketplace. And for comparison purposes we measure the competition's products in the same way and using the same criteria; these measurements provide information, but they don't dictate our recommendations, since they are no substitute for wisdom and judgment.

More important and much less tangible measures of a corporation's image have to do with how credible the corporation appears in times of crisis. What was Johnson & Johnson's image worth when it stood the company well during the Tylenol scare? Is there an accurate measure for the premium that the stockholders and management would receive in the event of the sale of the company, if the corporation's image is high? When rumors floated that Harley-Davidson might be purchased, many commentators thought the price would be higher than statistical analysis might warrant, because of the special image that the motorcycle manufacturer enjoyed.

In more everyday situations, a corporation's image can be measured in terms of the ease with which it attracts and keeps new employees and the willingness of its employees to "go the extra mile" in their work. Image is, as we have seen, an important factor in corporate culture. But textile giant J.P. Stevens didn't take this into account when it moved corporate headquarters from its factory area in the South to New York City; after this move, morale declined. Overhead costs were not the only item affected by the move. The corporate culture and spirit were affected, and eventually this led to the company losing its independence to an archrival—a terribly measurable event.

What Is the Book Value of Your Brands, And Who Knows It?

In Great Britain, accounting rules allow British companies to evaluate the worth of their brands, and to list those figures as assets on their books. There are standards and formulas by which the worth of a brand is judged, and the same standards are applied to every company, so that the brand values are not estimated in response to the pressures generated by a particular transaction, but rather on an ongoing and permanent basis. Perhaps this practice might have been expected in a country that cherishes long heritage and history. Not so on this side of the pond. Federal accounting regulations in the United States do not permit a similar practice for American-based companies. We can list a figure for "goodwill," though it is an arbitrary one, and consequently most companies do not set the dollar value of their goodwill very high—but we are completely enjoined from listing what may actually be some of our most important assets, the worth of our brands, and are also precluded from amortizing these assets for income tax purposes.

Perhaps you think the disparity is not much to get agitated about, a tempest in a teapot. Actually, this simple distinction between European and American accounting rules on brands has hurt us, and should be changed. And not just for the purposes of bookkeeping or fancy financial footwork.

Earlier in this book, I argued that brands may be the most valuable assets that a company possesses. As an asset, a brand has intrinsic value and great viability. For one thing, it is entirely proprietary. For another, it is irreplaceable, without real time limits on its use and impervious to technological change. Very likely this wonderful asset has been created subsequent to the time when our outdated accounting rules were brought into existence. Weston Anson, chairman of Trademark & Licensing Associates, a California consulting firm, recently wrote this provocative sentence in a technical article: "If the top ten US advertisers and consumer goods companies were to value their trademarks and brand names . . . they could probably add $25 billion to their balance sheet value."

It is axiomatic in business that top management gives its time and energy to items in direct proportion to the dollar value attached to them in the annual budget. *If brands are carried on the books at anything approaching their true value, the chances are that those who directly manage them will demand and obtain more attention than they are currently getting from the company's top executives.*

Let's consider for a moment what would happen if a company were to carry a brand on its books at a hefty and realistic value, say, in the millions of dollars.

In the apparel industry, there have recently been some cases in which the value of trademarks has been established. A ranking expert is James M. Cegelski, Senior Vice-President of Houlihan Lokey Howard & Zukin, a San Francisco firm; in a recent interview in *Women's Wear Daily* (April 4, 1990), Cegelski argues that income streams come from trademarks and can be assessed. For an employee buy-out of The Cherokee Group, his firm valued the Cherokee trademark at $35 million; in a refinancing of Aca Joe, then in Chapter 11, the court accepted the firm's valuation of $1.5 million for the Aca Joe trademark. Further, even after the Sasson and Robert Bruce companies had gone into bankruptcy, their trademarks "were sold for millions after the companies that established them were no longer around."

When Ya Comin' Back, Ryder?

Lippincott & Margulies was directly involved in one quite interesting valuation problem. In the 1970s, Ryder Systems had established itself as the dominant truck rental company among consumers who rented trucks. Founder Jim Ryder had carved out a particular niche in the trucking industry, a niche that very deliberately excluded involvement in the government-regulated industry segment that provided large customers with both trucks and drivers to haul loads across the nation. Jim

Ryder was simply not interested in getting into any business that was regulated—period. However, he was a flamboyant and promotion-minded entrepreneur, and the industry knew it. A company in the regulated sector, named PIE, that was looking to grow and to enhance its image, approached Ryder with a proposition that only in retrospect seems to have been a matter of smoke and mirrors. PIE proposed that it be allowed to rename their company Ryder/PIE; this would give it a higher level of awareness and an image of trustworthiness in the marketplace. On the other side of the highway, Ryder Systems would appear to be larger than it actually was. More vehicles would carry a version of the Ryder name without Ryder's having had to invest any additional funds, which at the time were in short supply. This appealed to Jim Ryder, and he let Ryder/PIE use his name and decorate its forty-four-foot-long trucks with a logo that included it.

At that time, Jim Ryder had no idea what his brand name was worth. He measured its value in terms of illusion—its potential for impressing others—and was not concerned with whether that impressiveness was real or deceptive; certainly he did not consider the possibility of measuring the value of his brand name in terms of cold cash. When the 1980s brought the impossible—deregulation of the entire trucking industry—Ryder Systems, now managed by a new team headed by the savvy, energetic Tony Burns, wanted to participate in *all* sectors of the trucking industry. The company acquired two trucking firms, ICCC and TransWestern, but because it had already abandoned the rights to the Ryder name in that particular sector, was forced to operate these two divisions under their current names. As knowledge of Ryder's entry into the new sector spread, so did confusion and conflict between Ryder Systems and Ryder/PIE. Further complications arose because Ryder/PIE services had deteriorated badly: because of the confusion in customers' minds about the two entities, Ryder Systems began to receive multiple complaints about "bad Ryder service," which was actually bad In Ryder/PIE service. Ryder was taking a bum rap, since its own operations were the best in the industry.

The new management recognized the great value of the Ryder franchise as a trucking image, and determined to buy back the exclusive rights to the Ryder name. A long and torturous negotiation between Ryder Systems and Ryder/PIE ensued, and the crux of the matter was determining what the Ryder name was worth. Without a widely accepted formula such as the British use, other figures had to be amassed—the out-of-pocket costs to repaint Ryder/PIE trucks, to change all the company's signs, to advertise a new identity, and to absorb all the other costs involved in a name-change program. Beyond these hard costs, the value of the intangibles also had to be established. While the final purchase price is privileged information, I can state that many, many millions were paid to Ryder/PIE by Ryder Systems to buy back the full rights to the name once traded because Ryder Systems had wanted to appear larger than it was. Jim Ryder, I am sure, was astonished to learn the true value of the identity he had earlier, and so blithely, given away.

Weston Anson suggests three basic tests of whether a brand has any real value: 1) "Does the name or trademark differentiate the product in the eyes of the user or the purchaser?" 2) "Would the name or trademark or logo have value to a competitor?" 3) "Would another company or competitor pay to buy—or to license—the brand name or trademark?" All assess worth by reference to the brand's performance in the marketplace.

What Am I Bid?

Consider the effect of the proposed accounting change on mergers or acquisitions. If the real price of a company that is for sale includes the correct dollar value of its brands, a higher price will be paid for the company. Or, I should say, an unduly low price or instance of overpayment based on a guess as to what a company's brands may truly be worth will become less likely. Did Philip Morris get a bargain when it bought the many brands marketed by Kraft Foods? Was RJR/Nabisco really only worth the original $17 billion offered by Ross Johnson and his cohorts, or was it truly worth the $26 billion paid by Henry Kravis and his associates? While the price was certainly influenced by a variety of factors, the $9 billion differential has at its root the inability of management to accurately estimate the true value of its branded assets. Analysts could only estimate the worth of the RJR/Nabisco and Kraft brands, for they were not listed on a balance sheet. However, a British firm was able to find and place a number on what the British retailer GrandMet bought when it acquired the Smirnoff brand from Heublein: 500 million pounds sterling.

Deeper Pockets?

In a global marketplace, since other countries permit the carrying of brands on the books at considerable value, U.S. companies are at a disadvantage. The practice allows a British company to make a higher bid for an American asset than a rival American company. For example, WPP knew that the actual value of its two new acquisitions, the J. Walter Thompson and Ogilvy & Mather advertising agencies, was greater than the book value, because the book value didn't take into account the prestige associated with the "brand names" of the agencies, names that had long been synonymous with creativity and excellence. WPP could then outbid rivals because, when the company sought financing for the takeover, it could show a British bank that according to its own value analyses—and bookkeeping that included figures for brands, or, in this instance, the prestige of the JWT and O&M shops—its balance sheet would look better, thereby enabling banks to justify extending greater credit to WPP, and bond investors to lend it more money for the purchase than could have been obtained by a less debt-laden company. The same thing happened when GrandMet bought Pillsbury; it paid a higher price than any American suitor would offer, because GrandMet valued Pillsbury more highly than the company was allowed to value itself! In a sense, Pillsbury didn't really know the value of its own brands, and didn't bother to figure it out, probably because its accounting firm said it would be unable to use the number for bookkeeping purposes. This is a classic example of tunnel vision, where concern for the numbers is the be-all and end-all of every analysis, the only dimension considered to be of use when developing policies and procedures or when making decisions. And here was a number that could have been useful—but wasn't figured. Tunnel vision in Pillsbury's case contributed to the management's inability to communicate the true value of the company to any white knights in order to prevent the takeover. Another consequence was the immediate unemployment of the old management soon after the takeover by GrandMet.

So, dear reader, try this simple but perhaps revelatory test. Until the day comes when accounting procedures here grow more enlightened, make out a new balance sheet for your company, one that lists its brands as definite assets of full and real value. See what this will do to your priorities, your allocations, and your annual expenditures for matters associated with keeping the brands prominent and viable, and what it will say about the overall health of your company. Does the change affect your debt ratio? Maybe the company is worth more than you used to think it was. Maybe there are some new strategies you'll want to try, now that the figures demonstrate where the real strength of the company lies . . . .

A Kinder, Gentler Budget

Fundamental to all budgeting activities is the wish to minimize surprises and to maximize that which can be predicted. A twenty-year lease is preferable under this principle to one of ten years, and an executive bonus pool based on a precise percentage of profits is preferable to one that slides about on whim. A corollary: permanent items in a budget are easier to plan for than one-time-only items. In fact, the more items that can be predicted, the more precise the budget and the more accurate will be the estimates of profit performance, this last a matter of great comfort to the board of directors and others interested in your company's performance.

All this is preamble to arguing that every corporate budget ought to have a line item in it for image management. I am chagrined to report that such a line item exists almost nowhere in the budgets of our major clients, but hope springs eternal.

There are three major reasons for having such a line. The first has to do with the expandability of certain items under stress and strain. When Perrier recalled its products, an extraordinary expense was incurred, one large enough to play havoc with any budget. But to deal with the reintroduction of the product, the company simply announced that it would increase its advertising expenditures from the previously agreed-to level of $6 million a year to $25 million. Since $6 million had already been budgeted, and the budget line was well established, it could be expanded. Note that a new $25 million didn't have to be found, just an additional $19 million. So: to make life somewhat easier, have some money in the budget that can be applied to image management in times of crisis, when unexpected efforts in communications are dictated.

The second reason: Money spent in calm times buys benefits that last for years. If you have invested in good image management practices for a considerable period of time—even at a modest level—you can be expected to have an image that is able to survive a direct hit such as the one taken by Perrier. If an emerging bottled-water firm suffered the same sort of injury, and had not had time to build up equity in its image, the bill would have been a lot more than $19 million.

The third reason: In a catastrophic year for the company, when expenses simply have to be cut, if you've spent time and money previously building up your image, you can cut back on that line in the budget and temporarily coast on the fruits of your long-term efforts.

A superior corporate image can be paid for, a bit at a time, through the sort of kinder, gentler budget that recognizes image as a basic contributing factor to the company's health; if this is done, the benefits are long-lasting.

An examination of the budget leads us directly toward a look at an even more fundamental aspect of a business: how it is organized.

CHAPTER 15

Corporate Parenting

These days, large corporations have many and sometimes disparate parts, and are faced with the important issue of deciding how the parts ought to fit into the whole, and how the parent should relate to its offspring, which take the form of divisions or subsidiaries. Analyzing the subject within a communications context is an excellent way to begin; for instance, consider what a particular executive's business card should say, and you get right to the heart of the matter. Does he work for a division of a giant corporation, or an independent unit? What's his title, and how does it fit into the entire corporate structure? The problems of corporate parenting are intensified when corporations acquire others that are not precisely in the same line of business as the acquirer has been. As with the nuclear family, things were simpler in the good old days of corporate parent-offspring relations.

Aligning for the Future

A classic case of communications being the key to parent-subsidiary relations is the story of the transformation of the old United Aircraft Company. When Harry Gray took over as CEO, there were several large problems to address. Because of its name, the company kept getting confused with United Airlines; some of the country's largest newspapers, on receiving advertisements from both of these companies, would run a United Airlines ad when it was supposed to be featuring one for United Aircraft. The perception of the company was also a problem in regard to how the company was viewed by financial analysts. Defense company analysts were assigned to follow the stock, and it moved in tandem with the defense budget. If there was an outbreak of peace, the stock went down, but if Khrushchev or Mao Zedong sounded particularly menacing, the stock went up. Beyond that, however, the problem was the way in which the divisions related to the parent company. United Aircraft consisted of such divisions as Pratt & Whitney, a manufacturer of what it billed as "Dependable Engines"; Sikorsky helicopters; Norden, a manufacturer of bombsights and other military devices; Carrier, a maker of air conditioners; Otis, a maker of elevators; and companies that made products that went into most American cars and trucks. All these divisions were strong, but the company was not deriving much benefit from their individual strengths in terms of its overall stock price; nor were the divisions being properly perceived by headquarters. Furthermore, new CEO Gray wanted to be able to expand the parent beyond its World War II–era base of companies, and to build a contemporary, world-class, technologically sophisticated manufacturing giant.

Lippincott & Margulies recommended a name and logo change to help reposition the company to the financial community, and a new nomenclature system that would link the divisions properly. Our choice for a name was United Technologies—the first word would keep the historic associations of the old name, and the second word would be more descriptive of the company's base. We also designed a new logo, called a "technology mark." It is a series of lines in the shape of a wheel, lines that go from slim and independent of one another to broad and contiguous as they round a circle in counterclockwise fashion. Some people saw this technology mark as the inside of a jet engine design, or as an abstraction of a gear—whatever, it certainly and unmistakably connoted high technology. Our further recommendation was that all of the subsidiary companies be linked to the parent through what we call an endorsement system: Pratt & Whitney, a United Technologies company. This concept responded both to the emotional and the pragmatic needs of the parent and subsidiaries. The subsidiaries could continue to benefit in the marketplace from the well-established name recognition, and the parent would benefit from the broader perception of it that would come as a consequence of being visibly and continuously linked with a number of diverse enterprises.

Gray liked the recommendations, and wanted to put them into place as soon as the board of directors' approval was secured. We suggested that he allow us to present them to the board of directors first, but Gray, a notoriously strong-willed leader, believed that the directors would approve anything he wanted to do, and in a board meeting tried to ram through the new name and design system. As we had feared, it was voted down. A CEO can alter almost anything about a company and receive great support from the board, but a name change, Lippincott & Margulies had long ago learned, is a subjective matter and one that often meets with resistance; time must be taken to form a consensus around an identity change recommendation. Acknowledging that it had been easier for him to get approval for a $100 million acquisition than to obtain permission to change the company's name, Gray then asked us to become more involved. We met with each board member separately, and also presented the concept formally to the board, offering explanations for the changes—for instance, our belief that the current name encouraged a misperception of the company's reality— and fielding questions that allowed us to elaborate on how we thought the new alignment would help.

In any event, the board did approve the new system, and United Aircraft became United Technologies. The logo went on everything from the company's stationery to its helipad, where it made an excellent target for craft as they were settling in to land. The subsidiaries were delighted with the nomenclature system, since it allowed them to be favorably linked to a now forward-looking parent company, while preserving their own heritage and integrity. People who had thought of Carrier air conditioners as a small company now understood that it also enjoyed the resources of a high technology conglomerate. As for the parent, the realignment allowed financial analysts to at last recognize the company's great scope independent of both the airline industry and the annual configurations of the government's defense budget. And Harry Gray had a new base from which to buy additional high technology companies.

Fashions in Corporate Structure

In the 1960s and 1970s, a conglomerate was the thing to be, a corporation whose divisions are associated only in the sense that they are linked financial entities. Although United

Technologies stressed the high technology similarities of its subsidiaries, many other conglomerates prided themselves on the disparity among their divisions. In this period corporations that had many dissimilar subsidiaries liked to say they were shielded from earnings declines because of their diversified income base. But they also said they were striving for synergy among their components, to achieve the status wherein the whole would be greater than the sum of its parts. (That these statements contradicted one another did not faze the proponents of the conglomerate structure.) The role of the conglomerate corporate parent was to finance the subsidiaries, and make acquisitions and divestitures. For the corporation's stockholders, the parent acted almost as a mutual fund, investing in a variety of businesses to insure a steady stream of profits that would be insulated from the inevitable swings of the business cycles that affect all corporations. The responsibility of each division was to "do its thing," and to pass its profits on to the parent—no more, no less. Any relationships between sister companies were, at best, informal. This parenting posture could be characterized as cold, impersonal, and focused solely on the bottom line. The ties that bound the parts together were made of a substance some consider thicker than blood—money—and a sense of family was entirely absent. In such an environment, the image of the parent was important only to the financial community, and managements became convinced that the only thing that mattered to that community was the conglomerate's profit and loss, its balance sheet.

Nowadays we hear a lot about corporations striving in the opposite direction, to become "pure plays" in transportation, entertainment, or finance—not all three at once. Gulf & Western began the decade of the 1980s as a prototypical conglomerate with a disparate portfolio of unconnected businesses. It recently became a pure play in the information/entertainment field by spinning off its successful financial services division and adopting the name Paramount Communications to embody and signal the change. Under this new structure, Simon & Schuster, the publishing division, is allowed to be a part of Paramount, yet for marketing purposes maintains its separate identity, without the publishing company or its employees—or its authors, or even the buyers of its books—having the feeling that S&S is submerged and unknown in an omnidirectional conglomerate.

Since fashions change, and change constantly—is it not a definition of fashion to alter continually as time passes?—something must remain constant in a corporation's relationships with the world, and that constant is communications practices. Communications can help to define the role the parent is to play, to foster its relationships (parent to subsidiary, division to division), and to inform all audiences, internal and external, about the company's chosen configuration and missions. Among the central, and quite vexing, problems of parent-offspring relations are how to convince the various audiences of the correctness and the adequacy of the chosen parental role, and how to show the financial audience that the chosen role will add value to the subsidiaries and to the corporation as a whole.

Here are some of the challenges, as they might show up in questions raised by today's volatile corporate parenting climate:

• Does an assembly-line worker toil for Chevrolet or for General Motors? Ask him! If he doesn't give the right answer—GM—then the leadership role of the parent is being wrongly perceived by an important "internal" audience, the line work force.

• How do you sustain the loyalty of your employees if one day you're known as AT&T and the next you're NYNEX? To fail to inspire continued loyalty is to fail as a parent.

• How do the managers of American Can feel now that the company is owned by the French government and they are, in a sense, working for a foreign power? Conversely, do employees of American Motors resent the recent change of ownership that now has placed the company in the hands of what was once its archrival, Chrysler?

• How does a synergy-driven conglomerate convince the financial community that synergy is an important aspect of the company's future plans, when synergy is frequently viewed with skepticism by sophisticated investors? Saatchi & Saatchi announced that it would buy a bank because it felt it would be an ideal synergistic complement to its advertising and consulting services; its announcement was viewed with derision. Yet when Philip Morris acquired Kraft Foods, the purchase was applauded for its synergistic excellence. How should a parent define and communicate its particular concept of synergy?

• How does the Ogilvy & Mather executive feel about working for a man whom David Ogilvy, the founder of the company and the idol of many people in the advertising business, had publicly labeled "an odious little jerk"? How can the new owner of O&M ever bind the old employees to him?

• How does an employee feel when he learns that the division for which he works is on the block, and may be sold by the parent to raise cash? Will the proposed sale cause him to bolt immediately (and by doing so perhaps lower the true worth of the division)?

- What do you do with a parent that is simply a bad parent? One whose strategy is severely damaging, if not actually killing, its divisions, and whose misguided instincts may soon prove fatal?

This last problem is the extreme one, and may even be the simplest to resolve, if we are speaking only from a communications standpoint. Case in point: the Campeau Corporation.

Gilt by Association

In 1987, the Campeau Corporation was accurately perceived by those people who knew of it as an aggressive and adroit Canadian real estate development company; no more, no less. Founder Robert Campeau must have become dissatisfied with this particular image, and while I do not suggest that his subsequent foolishness was motivated exclusively by a desire to change his company's image (and, because of the shared name, his own image as well), image was clearly a consideration in all that followed. Putting aside such obviously crucial factors as finances, business strategy, the art of persuasion, and the greed of the investment bankers, not to mention business cycles and the cruelty of fate—all of the vital determinants of later corporate health notwithstanding—the Campeau Corporation had an unsatisfactory image. It had low visibility, low stature, low prestige, and low glamour. To convert the lows to highs, Campeau embarked on a plan to buy jewel images and have them reflect well on the parent organization, a plan one might call "gilt by association." Wonderful images of such stores as Bloomingdale's, Jordan Marsh, Burdine's, Abraham & Straus, Rich's, Lazarus, and Bon Marche were perfect vehicles to elevate the plebeian Campeau the real estate company into the patrician Campeau the glamorous department store holding company. It was if a poor immigrant had adopted a bunch of children who happened to be graduates of Harvard, Princeton, Yale, Columbia, and Brown, and expected to be perceived as an Ivy Leaguer by virtue of his children's accomplishments.

But buying instant prestige, for Campeau, was a dangerous game, and an extremely expensive one. If the basic strategy was not flawed from a perspective of pure image, it was certainly fatally flawed from a financial standpoint. Campeau could not afford his luxuries. Perhaps he overpaid, perhaps he was overleveraged, perhaps he was just at the wrong moment in the business cycle—people will be arguing about that for years, but the end result was that he ran out of money long before the glitter could gild his parent corporation's lily.

So much for the Campeau corporate parent. What could the children do to survive? After all, they were the entities with true value, the owners of hard-earned, well-polished images. Their performances proved as sterling after they were adopted as before. My first recommendation to them, after the incapacitation of their parent, would have been to find a new chairman and a new name, to dissociate the real assets from the failure in every way possible. And that's what happened. G. William Miller was announced as chairman, and the stores took a new/old name, Federated Stores, Inc. In essence, then, the children divorced the parent, sending out a signal—a new boss and a new name—to everyone. All hope that reality in the form of better corporate health will follow the signal.

As can be seen from the Campeau case, problems such as the examples suggest cannot be ignored, because they won't simply wither away. The penalty for underreaction can be severe, and in all cases of neglect, the rot will spread rapidly and affect the very foundation of the company, its relationship with its employees. Let me now point out that most managers and corporate owners merely pay lip service to the notion that employee morale has anything to do with profit and loss—but consistently successful companies know that good relations between employees and the corporate parent are of the essence. There are many examples of catastrophes that can occur if this priceless asset is overlooked, taken for granted, or—worse—discounted to the point that employees are deliberately antagonized and become willing to obstruct management's plans.

In order to develop communications plans that will help a company be a good parent, it's necessary first for the company to have a clear mission statement—a description of what sort of structure you have in place (or are striving for). Only then can it be communicated loudly and clearly to your audiences. Many companies haven't properly defined their roles. Others have defined the roles, but haven't pursued communications practices that let their audiences know of those roles and the consequences of their particular alignments. As a primer, then, I present three modern models for parents, as a communications expert views them.

Keep Your Logo On

At one end of the scale is the venerable General Electric company, the prime example of an extreme, monolithic communications model. Virtually all of GE's divisions are under tight direction from headquarters and have very little in the way of separate identities. The identity of acquired companies is replaced as soon as is practical by the GE identity. The GE name is placed on every company product, from locomotives to light bulbs. Why, after acquiring RCA, GE even changed the hallowed RCA sign on the 30 Rockefeller Plaza building in Manhattan, so that the red letters would have the proper corporate logo. (And it took a lot of flak for doing so; some people said they'd *never* refer to the skyscraper as "the GE building," but in ten years, I promise you they will do just that. I say this through gritted teeth, because GE's decision hurt one of my favorite sales aids. Lippincott & Margulies designed the RCA logo and went to great pains to adapt it for the sign on the 30 Rockefeller Plaza building. For years, from the window of my office I was able to dramatically point to it as a living example of our work. *Sic transit gloria mundi.*) By placing its logo on all products, GE is communicating that the company is a proactive parent and the hands-on manager of everything it touches. There is no room for ambiguity in the parental role, nor, thanks to its communications, how that role is perceived by the internal or external audiences. All advertising is controlled by the parent and features the parent prominently, and the product as the fruit of the GE tree.

Call Me When You Need Me

In the middle of the road (as far as parent-offspring structures go) is American Express. American Express imposes its corporate identity as a primary marketing identity on some of the products it markets, principally those traditional ones that have to do with its travel services—Travelers Cheques and credit cards—but not on all of them. For instance, its important division IDS is viewed as independent, and so, to a degree, was Shearson Lehman Hutton, the brokerage house. Advertisements for Shearson Lehman generally carry a line identifying the brokerage as an American Express division, but it's a soft-pedaled line. The theory seems to be that if identification with the corporate parent will assist the division—say, by suggesting that American Express's financial might stands behind Shearson Lehman—announce it; if not, soft-pedal that identification. We didn't design the blue box as a container for all of American

Express's divisions, we designed it to be reflective of the parent. A distinct advantage of this communications architecture is the flexibility it provides to the parent. During good times, Shearson was allowed to bask in the glory of its own performance and profitability, and to loudly trumpet its contributions to American Express's bottom line—but when trouble loomed American Express could step in as a supportive parent who, in the extreme case of Shearson, could become an absolute savior. When Shearson has been once again turned around, the former relationship of subsidiary with parent—"Call me when you need me"—can be revived.

Anything and Everything

Off on the other end of the scale from the GE, all-things-are-part-of-the-parent model, is Primerica, a modern conglomerate. Primerica describes itself in its annual reports and other communications as a company whose mission (and obligation) is to maximize return on equity to its stockholders. Period! What business is Primerica in? Anything and everything. This is not so shocking, since the only real requirement is for a corporation to elucidate and attempt to fulfill its primary mission, whatever it decides that should be.

After the conglomerate had sold its packaging division together with the name American Can, Chairman Gerry Tsai wanted an identity system that would allow the company to buy and sell any sort of subsidiaries at will, and we helped provide it, together with the name Primerica. Even after Tsai sold the entire company to Sandy Weill's company (formerly known as Commercial Credit), Tsai's policy was perpetuated: the company has not wanted to be bound to a specific line of business by becoming overly associated with the fields of endeavor of any of its subsidiaries. Following this strategy, Primerica's communications always emphasize that the corporate parent is separate from its component parts and subsidiaries. One can't buy a Primerica anything—except a share of its stock. Employees will tell you they work for Smith Barney, or Fingerhut, or Sam Goody's, not for Primerica. The parent company identity is something aimed exclusively at the financial community, while the subsidiary identities are for marketing purposes.

Nourishing the Institutional Image

Part of the reason the Primerica and Smith Barney/Fingerhut/Goody's images "work" is that they are all aggressively nourished. In the service industry arena, this is not always the case.

Nourishing an image of a corporation is always important, but more easily done for manufacturing corporations than for those who provide services. It has always intrigued me that manufacturers will advertise and otherwise enrich their trademarks, but that service company trademarks are not viewed and treated in the same way, even though the need to indelibly impress a service company's identity on its audiences may be as great as that of a manufacturer. Part of the reason for this reluctance comes from professional cultural difficulties: doctors, lawyers, and accountants who call attention to themselves by advertising or promotion are seen as downscale, not sufficiently in demand, stuck with appealing to the masses and not to discerning clients.

Other difficulties: Whereas a product can be seen and touched, and is therefore more easily understood and tangible to its consumers, a service is less tangible and less readily visible. Then, too, products can be described with precision, including a description of their salient features and the price; often services do not seem to have similar features, and their prices may not seem to bear a discernible relation to underlying costs. Also, while products are often purchased on impulse, services are not—one buys them in response to a specific need, such as wanting to own or sell a stock, or requiring legal assistance in making a will.

But the necessity to nourish it exists. Consider what would have happened to Fidelity Investments at the moment of a key retirement if it had not nourished its image. When Fidelity Investments' resident genius Peter Lynch unexpectedly decided to pick roses instead of stocks, some people expected disaster— that the Magellan Fund Lynch had headed for so many years would go south, and that its fall might impair the entire Fidelity empire. That didn't happen. The strength of Fidelity was sufficient to avoid an avalanche of redemptions in the Magellan Fund. Fidelity had nourished its institutional image, and because of that was able to survive the retirement/departure of a key individual—indeed, one whose name was widely known.

How can other service institutions insure their own future endurance?

The first necessity is to break down the attitudes of the management of service companies that equate image-shaping activities with advertising, and that decry image shaping because they think it will mean having to advertise. As I've pointed out earlier, there are many components of a corporation's image, and not all are dependent upon advertising.

If a company has come to the understanding that nourishing the institution is an important primary task, many possibilities fall into line. Clients must be made to feel they are doing business with the institution, rather than with the individual. Lawyer Arthur Liman was profiled in a major newspaper after his defense of Michael Milken—and it was not until very deep in the article that a reader could find the name of the law firm of which Liman is only one of many partners: Paul, Weiss, Rifkind, Wharton & Garrison. In this instance, the individual is being celebrated to the exclusion of the institution.

I suppose that's somewhat inevitable with regard to big-time lawyers, but on a less dramatic level, think of the great number of insurance and stock salespeople running around nurturing their own personal images through intense selling efforts—and wondering if they need to stay with their institutions anymore. If a star salesperson quits his or her employer, will that star be able to take the business along, because the client bought the salesperson and not the organization he or she represented? This is the big danger for service institutions, something often left unspoken but always palpable: that the employee can walk off with business that rightly belongs to the company.

To make clients feel they are doing business with the company and not the individual, business cards, for example, should display the institution's name more prominently than the individual's. We've recommended this simple practice a number of times, and have had many arguments about it, howls of protest from stars and would-be stars; their very objections show why the practice is necessary.

Beyond such relatively trivial items, the company needs to have a consistent communications policy to reinforce the institutional image: newsletters, capability brochures, press releases—in fact, all external expressions. Sometimes, in our research, we call up service companies and ask for a current capability brochure, and are amazed to learn that many companies don't even bother to keep these up-to-date, a practice that encourages clients to shop around and inevitably to rely on individuals for the latest information.

The closer a service purveyor is to the public consumer, the higher the awareness of the need for image-building activities. This principle can be seen in action in the story of the transformation of a company originally called Extendicare. When the company decided to stop being a chain of nursing homes, and to go into the business of providing the general public with hospital care, a name change was essential. After all, a hospital with a name that implies that patients will be staying for extended periods of time is not likely to appeal to that broad segment of the general populace that dislikes staying in hospitals. Lippincott & Margulies persuaded Extendicare that while a new name was a crucial element in its communications, it needed to shape an entire image to influence the public. It needed to explain its unique business concept, to run a hospital as a profitable business, to both the financial community and the medical profession. We suggested the name Humana, one that came close to preempting the position of humanitarian care for its hospitals; beyond that, we developed a discipline that linked Humana to everything that the corporation runs. The Audubon Hospital became the Humana Audubon; its health insurance company, developed a few years later, was introduced as Humana Care Plus. In short order, the name Humana was widely recognized, and the discipline of naming each of the services under the Humana umbrella allowed each of the services to be institutionalized more easily than they would otherwise have been. In sum, the name treated these services as if they were products.

Similarly, stock brokerage houses that deal directly with the public—Merrill Lynch, for instance—work hard to send out advertising, brochures, newsletters, and the like, and do everything they can to burnish the image of "the bull." They do so to establish in the minds of the customers a value-added dimension to their services. In contrast, investment banking houses, much farther removed from selling to the masses, are less convinced of this necessity. I recall a conversation with John Gutfreund of Salomon Brothers, just before the October 1987 stock crash; when he asked me what the difference was between his firm and its competitor Goldman Sachs, I replied that while Salomon had an image of being a collection of very bright stars, Goldman was perceived as an institution in which some stars resided, and that the latter image was preferable to the former. The institution was more long-term effective, able to withstand the brutal swings of a cyclical industry. Individuals might come and go, but institutions can last forever.

Long term: that's the key, here. To burnish an institution is a long-term effort, and one to which resources must be allocated on a continuing basis. For instance, talent must be recruited and developed to have the attitude that each person is a member of a team that works for the company, rather than being developed as an individual entrepreneur. After all, employees are also walking, talking media for the company's image, and in many instances are the major conveyers of that image. A demonstrated willingness to invest in employees—acting as if you care about them and their future—may entertain the realistic hope of inspiring their loyalty. Someone who has been trained by an organization, and brought along (often at substantial cost) to the point where he or she is enabled to become very valuable, will feel appreciated, and will have a greater likelihood of staying with the organization (and not walking off with its clients) than the person who is simply hired, put on straight commission, and told to sink or swim. Obviously, investing time and effort in employees is a dollar risk, but one that invariably pays off in the long run. Again, as I have often emphasized, image is a long-term activity.

Beyond A, B, and C

There is an evolving model that some forward-looking companies are pursuing, among them several that are our current clients, and whom we are advising in this process, because parent-offspring relations is an important factor in the management of corporate image. These companies are working their way toward a partnership between parent and offspring. This concept is rooted in the notion that each offspring has its own value, culture, and strengths that are worth nourishing; further, that this nourishment can be better achieved because of the particular resources of the parent. Both parent and child need one another, and both benefit equally as a result of the relationship. Implicit in such a concept is the autonomy of each division, as well as carefully proscribed limits on parent involvement, and what amounts to an almost complete disregard for the other siblings within the family. The idea of synergy—so difficult to attain—is replaced by a genuine attempt to maximize existing asset strategy. For example, Lippincott & Margulies is a member of the Marsh & McLennan Companies. We have little synergy with any of our sister companies, since we don't sell insurance, or mutual funds, or environmental or economic consulting services. We each maintain significant positions in our respective marketplaces, yet all are nurtured by a common parent, and are encouraged to develop our own individual images to the fullest while maintaining cordial and informal relationships with one another. Can an individual be both an American and a New Yorker? An American and a Catholic? Of course one can, and a similar dual identification (of belonging at once to the larger and to a smaller group) is what a "corporate parent partnership" seeks to promulgate.

Parenting and the External Audience

No matter which model your company chooses, problems are bound to arise because of the relationship of parent and offspring. Many have to do with the morale of employees, but others have to do with external audiences. Most of these can be addressed by making sure that the parent's image is well regarded. Consider the case of Philip Morris. As a purveyor of tobacco, a product that stands accused of causing cancer in millions, the company is often labeled as a merchandiser of products that can kill. On the other hand, Philip Morris has many divisions that market products such as Jell-O and Philadelphia Cream Cheese that are clearly not in the same category as the tobacco products. How can the company position itself so that it is not overly perceived as a tobacco company? Should the Philip Morris logo go on a packet of Jell-O? Almost everyone would conclude that such an identification would hurt the sales of Jell-O.

Philip Morris's 1989 annual report shows how it addressed the problem of corporate relations with subsidiaries, and at the same time sends an important message to the financial community following its stock price and performance. The report features a double-page spread that consists of three equal-sized photos, one of cigarette products, a second of an enormous range of food packages, a third of Miller beer; the reader of the report might well conclude that all three parts contribute equally to the company's makeup and profits, when in fact while tobacco products make up less than 40 percent of its revenues, more than 65 percent of the company's profits come from tobacco. Similarly, the company describes itself as being in "consumer goods" and says further that it is "balanced." Having said such things for several years, Philip Morris has had the satisfaction of being followed by analysts other than those assigned to tobacco, and evaluated differently from the stocks in those other groups. While Philip Morris's stock multiple is lower than the multiples for pure consumer packaged goods companies, mostly because of its tobacco image as well as a belief that it is vulnerable to litigation against its cigarette products, the company itself is at least accurately understood. In recent years, because of the spectacular performance of Philip Morris, it has replaced IBM as the bellwether blue-chip "industrial" listed on the NYSE.

Similarly, as one of the largest producers of armaments

and material for warfare, General Electric is actually a major defense contractor. Unwilling to be labeled solely as such—and perhaps fearful that being in the armaments business would not help the sales of GE washing machines or locomotives—GE some time ago decided to adopt an aggressive advertising campaign of "We bring good things to life." This skillful advertising campaign enhanced the burnishing of the parent's image through monolithic corporate identity practices.

Management clearly understood that visibility plays a major role in shaping the public's perception. If the GE identity became a permanent fixture in millions of homes through its refrigerators, irons, coffee makers, and other helpful appliances, the company believed, GE would eventually be perceived as a friend who is allowed and even invited into the home. The advertising campaign reinforced the wholesome helpfulness of GE products, and obscured and overshadowed GE's more lethal and less attractive product lines in the public's mind.

GE may simply have been following the lead of Du Pont in this matter. The chemical company was once known as "the merchant of death," and a book was written about Du Pont that bore that title; now, after many years, Du Pont has become perceived as congruent with its slogan that the company makes products "for good living." Its identity is also highly and permanently visible on a host of garments and quite wonderful consumer products.

Benefits from Good Parental Relations

The parent's communications policy must not be perceived by its audiences as solely for the benefit of the parent. The parent cannot appear arbitrary, eternally imposing edicts that may be opposed from below or that compromise the autonomy of the subsidiary or those in it. The parent's usefulness and role need to be both understood and welcomed within the ranks. If the parent is perceived as only an expensive and largely unnecessary overhead cost to the subsidiaries, the parent will never get from them the cooperation and enthusiasm it desperately requires.

I recall one instance where a divisional manager in a company that had just reorganized was adamant that the paychecks for his division not be changed so that they would come from the corporate parent—because he believed that if that happened, he'd lose power over his own troops. The corporate parent could simply have imposed its will, but that would have gained it little but a disgruntled line manager; some hand-holding and polite persuasion was necessary to bring this "child" to the understanding of the greater objective—corporate unity—that required him to give up one of the symbols of his division's independence. He needed to feel that his turf was not being reduced, but, rather, that his importance was being enhanced.

That sort of hand-holding was necessary to achieve consensus for the change in structure. I mention it, also, because a corporate parent's communications about relations with subsidiaries must reflect operational reality; that is, a parent can't say, for instance, that it is a partner and then fail to act as a sensitive partner would.

If, as with most corporate parents, the owning company does not actually create revenue, the parent does establish priorities for the application and uses of the revenues created by its divisions. Further, the parent gives the divisions access to capital that they would not otherwise enjoy; extends to their managers a better quality of stock options than if the divisions stood alone; and—not the least important of its benefits—with its many sources of revenue functions as a big security blanket for a division that has fallen upon bad times. In most good corporate headquarters-and-divisions relationships, then, there exists a mutuality of complementary self-interests. Well-planned and well-executed communication practices can achieve understanding of this reality on both sides of the parent-offspring gap.

But what about the greatest purveyors of corporate headaches in our era, corporate takeovers, mergers, and acquisitions? The image perspective can give us important information to help understand the dynamics of the company under siege.

CHAPTER 16

Image and Takeovers

The cardinal factor in takeovers is the financial health of the target companies, and the best defense is and always will be a terrific balance sheet, solid earnings records, a good profit margin, and—perhaps above all—a high multiple on stock price. These are the items generally used to calculate the financial formula that becomes the essence and language of the deal. Without in any way diminishing the central importance of financial factors, I want to point out that there are other issues involved in mergers and acquisitions, and that many of them fall under the heading of image. A cartoon reprinted in the *New York Times* purported to show a boardroom where the merger of Disney and the late Jim Henson's enterprises was being discussed; the Henson Muppet characters—Kermit the Frog et al.—said they were all for the deal, so long as they wouldn't have to wear those funny Mickey Mouse hats with ears.

I find this an excellent formulation of the notion that image is quite an important factor in takeovers, mergers, and acquisitions; in this instance, it embodies the idea that even the symbols of companies—thought to be totally unconnected to the balance sheet—can be deal breakers. No matter how seductive a potential deal may be to both sides, issues of appearance and pride can be just as significant in affecting the outcome of negotiations.

The foremost expert in the field of managing the image

and public relations component of mergers and acquisitions is Gershon Kekst of Kekst & Company, who has been involved in virtually every single major acquisition—both hostile and friendly—that has taken place in American industry in the past twenty years. According to Kekst, image is viewed differently by the two major types of takeover people.

Raiders

The first type, who rely almost exclusively on financial data to make their assessments of target companies, are wheeler-dealers interested primarily in immediate financial gain. They assign no value at all to image when considering an acquisition or merger. Since they consider unquantifiable such assets as image, corporate culture, and company history, they have no interest in these factors. What they are really seeking to buy are pieces of paper that describe mathematical relationships. And, more frequently than not, they buy with an intent to resell as quickly as possible, for maximum profit.

Now there is nothing illegal or immoral about wheeler-dealer types. They make no bones about their aims, and in many instances serve a useful function by replacing tired and under-performing assets with dynamic and productive ones. Financially driven opportunists, they are for the most part short-term thinkers whose principal motive to is build their own net worth as rapidly as possible. If, along the way, their interests coincide with those of the companies they own, so much the better for the companies; if those interests don't coincide, it is the company that suffers, because there is no question in the wheeler-dealer's mind as to whose interests come first. While some may feel that this attitude is quintessentially capitalistic and therefore justifiable—and many of these types have a habit of wrapping themselves in the flag when anyone questions their motives or methods—my own belief is that such behavior, when taken to extremes, weakens our entire capitalist structure. To think solely about the short term is to undercut long-term economic health; I prefer concern with the long term, and image is, as I have argued, an asset that cannot be created overnight and must be carefully nourished and nurtured for years to reach its full worth.

Despite the headlines, not all acquirers are destroyers, and only a minor fraction of mergers and acquisitions are hostile. But when acquisitions are undertaken with the obvious intent of later breaking up the target company, it is a certainty that the image of the target company is being valued by the raider at close to zero. T. Boone Pickens stated publicly his intent to break up Gulf Oil if he was successful in acquiring it. Could the value of Gulf as a brand name for gasoline possibly have been of any significance to Pickens in his calculations? I think not. Similarly, I am sure that William Farley had no notion of the effect on the company's executives of his announcement that he would bust up West Point Pepperell. As it turned out, his miscalculation cost him dearly.

The point is that a "breakup artist" calculates the value of the target company dead, rather than alive, and this misses the true value of a company, which may reside as much in the spirit of its employees as it does in the contents of its balance sheet. Would you drive a newly acquired car on two wheels if it ought to have four?

Developers

The second type are what might be called strategic buyers. They make acquisitions aimed at strengthening the core of the acquirer's business. By nature and inclination, they assign a much higher value to image. The example here is Ford's recent purchase of Jaguar, at an inflated price that can be explained only in terms of Ford buying a luxury image for an entry into the luxury-car segment of the industry that Jaguar has long cultivated and embodied. Such a purchase is certainly made for the long haul, and clearly not for the short-term breakup value of the acquired company. In fact, putting aside the obvious synergies and other benefits of the Ford-Jaguar union, it can be argued that Ford paid $3 billion just for an image, albeit one that took decades to cultivate, and one that Ford undoubtedly expects will be of enduring value if properly managed.

Sony's purchase of Columbia Pictures and subsequent acquisition of the successful individual producers Jon Peters and Peter Guber provides another example of a classic strategic purchase. Despite the high price paid, Sony must have no intention of dismantling Columbia; rather, Sony needs to have Columbia flourish for the sake of both the corporate parent and child, and evidently understands that this will happen only if Columbia is allowed to do what it does best, entertain the public. Columbia was purchased, in part, because of its image; knowing that, we can anticipate that Sony will follow up the purchase by nourishing its new studio properly in order to continue to build it for the long term.

Interestingly, the raider-type buyer can sometimes become the strategic-type manager—and vice versa. Consider Carl Icahn, who was dubbed the prototypical bad-guy when he began his takeover of TWA. In more recent times, Icahn was looked upon as a potential white knight to save Eastern Airlines from the unwanted embrace of Texas Air. And Gulf Oil thought it was better to be taken over by Chevron than by raider T. Boone Pickens; but since the purchase by Chevron, the Gulf brands are all disappearing anyway, so maybe Chevron as strategic buyer has a bit of the buccaneer in its makeup.

Myths About Image in Takeovers, or, Keep Your Facts Clean

Kekst defines image as "the element that differentiates one company from another, and that defines an expectation in the audience." Since takeovers encompass how companies perceive one another, how boards of directors and shareholders react to tender offers, as well as how the general public perceives the whole affair, this insistence on the importance of the "audience" pinpoints a crucial factor. In the past, three myths surrounded the subject of image in regard to takeovers.

1) That the image of great size, or of a company being a member of the establishment, was in itself a defense against being acquired. RJR/Nabisco was the thirteenth largest company in the United States when it was essentially taken over from its shareholders and management. U.S. Steel, which became USX, was the quintessential giant, but even after restructuring remains under threat of losing its corporate independence, as does Chevron, one of the largest energy companies in the world. Both have been rumored as targets. So much for sheer size being a defense against an unwanted takeover. General Motors, watch out!

2) That a company's stature within the establishment, together with its traditions of loyalty and strong corporate culture, were sufficient barriers to discourage unwelcome suitors. CBS had always been perceived as the "class act" of the broadcasting industry. It had all the requisite elements to be an unchangeable part of the American establishment: quality programming, a news department second to none, management stability, and a style that ennobled all it touched. Historic single-minded dedication to these qualities did not protect CBS from hostile attack by a very different breed of broadcast cat, Ted Turner. While Turner failed for financial reasons in his bid to buy CBS, the company's establishment posture did not deter Laurence Tisch from making his successful takeover of the company. Similarly, while Gillette continues to be able to stay out of the grip of would-be raiders Ron Perelman, Conniston Partners, and others, it is certainly not the Boston Brahmin heritage of the company that is its best defense; good financial performance and crafty legal maneuvers are the major shields for Gillette.

3) That nice people don't do takeovers, and, therefore, that the establishment would band together and reject anyone who tried to spearhead a hostile takeover. There's some substance to this myth, but it's fading. When Saul Steinberg tried to take over Chemical Bank, he failed, and was left, in his own words, "bloodied but unbowed." He's still a big player. And Kohlberg, Kravis and Roberts have an explicit understanding with their investors (America's premier pension insurance funds) that their money will never be used for a hostile takeover. Certainly the most astounding recent example of the myth was a judge's ruling in favor of Time, Inc.'s board that upheld the right of that board to dismiss out of hand an offer to shareholders of $180 per share from Paramount Communications when all that management was offering was about $130 a share.

As these examples show, the myth is still in flower, but fading. More usual these days is the experience of Carl Icahn. A raider not generally considered in the nice-guy category, Icahn never had difficulty raising money from establishment figures in his various takeover fights. Fred Hartley, whose Unocal Corporation was under attack by T. Boone Pickens, had to sue his own bank, Security Pacific, to arrest its financing of Pickens' hostile takeover attempt. If politics makes strange bedfellows, so do sky-high commissions, fees, and interest rates.

If it is clear that those sorts of defenses against takeovers are no longer viable, what sorts of image-related defenses can be mounted?

Cast the First Stone . . . ?

When American Express tried to take over the venerable publishing firm of McGraw-Hill, there was a furor. McGraw-Hill didn't want to be acquired, but was in difficult financial straits and did not have enough resources to fight off the would-be acquirer in traditional ways. But some research revealed that a man named Roger Morely was a member of the boards of both companies. McGraw-Hill then publicly accused Morely of violating his professional integrity (at the direction of American Express) by having passed to American Express privileged information about McGraw-Hill. Now in another industry, this sort of thing might have gone unremarked, but the publishing company mounted an interesting argument in regard to Morely. It contended that because a publishing company has at its core its reputation for integrity, if that reputation were to become tarnished, authors would take their books to other publishing houses, booksellers would not deal with the publisher, and so on. Therefore, said McGraw-Hill, it could not allow an "unethical" company such as American Express—unethical in McGraw-Hill's eyes because American Express had let Morely pass information—take over the old publisher, because if it did, the whole house of business cards based on integrity would collapse. Faced with this defense, American Express withdrew its offer, and today McGraw-Hill remains independent.

The Cult of Loyalty

Harley-Davidson has for some years been the sole remaining manufacturer of motorcycles in the United States; the others all went out of business as a result of competition from foreign manufacturers such as Honda. The reason that Harley had survived was that it made a good product; more than that, though, Harley had developed a cultlike status with both its dealers and the ultimate buyers and owners of the motorcycles. Knowing this sort of veneration was good for business, Harley had always encouraged it.

Despite Harley's success, competition from overseas had made some inroads, and the company was not as strong as it once had been. When an independent corporate raider made some noise about taking it over, management initially began to get its ducks in a row.

Through astute research, Harley learned what bank was financing the raider. As it happened, this bank had also made some large loans to Harley, so the president of the motorcycle manufacturer called the bank's senior officer and asked for a meeting. At that meeting in the bank's Manhattan headquarters, the Harley man informed the banker that Harley's customers and dealers were intensely loyal, and that if the bank continued to back the raider, he was going to put out the call for Harley's customers to come down to the bank's buildings and gun their loud motorcycle motors as they rode around and around it, until the bank agreed to cut off the raider's line of credit. Shortly, the raider found himself without his major financial backer, and Harley, too, remains independent. Customer loyalty—an image issue—proved to be a hell of a defense.

Blood and Rubber

Goodyear Tire and Rubber of Akron, Ohio, another venerable company, had been in the doldrums. Although the company was well respected and considered a benevolent employer, its stock was a poor performer, and Goodyear had become only marginally profitable. It was clear to many analysts that the company needed radical restructuring. Jobs would have to be eliminated, blood would have to be spilled; but for several years the management had resisted making these moves.

Sir James Goldsmith of France and Great Britain is a renowned dismantler of companies, and he decided that Goodyear was a proper target for his attention. Faced with an attempt by Goldsmith to buy the company, Goodyear mobilized. Knowing that it could not make a proper defense based on its numbers, it chose instead to rally its own employees and to get state legislatures and the Congress on its side. Goodyear painted Goldsmith as a vicious foreigner who would dismantle an American institution. Management argued that yes, changes needed to be made, but that Goodyear ought to make them and they should not be imposed by an outsider. The battle raged for some time. Goldsmith even testified before Congress, giving what was called some of the bluntest testimony ever heard in its halls on a business matter; he argued that if the company were not made competitive through radical restructuring, it would go under.

Goodyear's campaign was effective. Laws were passed that would have made it impossible for Goldsmith to control the business, and eventually the raider turned his attention to other targets. After that, Goodyear management did pretty much what Goldsmith would have done had he been successful—instituted massive layoffs, plant closings, rollbacks in wages and benefits for unionized workers, etc. But this was found acceptable by the work force and by the American public, because it was done from within. By relying on the built-up goodwill of its employees and of the community—another clear image issue—Goodyear retained its independence.

How Will It Look?

When Shearson Lehman Hutton teamed up with RJR/ Nabisco CEO Ross Johnson in an attempt to engineer an LBO of the food-and-tobacco giant, they felt that their efforts would be successful if they were able to convince the public that they were acting for the benefit of their stockholders. Ross Johnson asserted to his board of directors that he had done everything within his power to boost the price of the company's languishing stock, and had failed; therefore, he said, the only alternative was to simply buy out the shareholders for a price reflective of the true value of the company. His argument was that the stockholders would win, and that the intrepid team behind the LBO would win eventually if it improved the performance of the company after the change. "Take the money and run, shareholders," he was saying, in effect, "and we'll take all the future risks."

Unfortunately for Johnson and Shearson Lehman, things weren't that simple. In their arrogance, they underestimated the analytical abilities of several groups—the breakup artists, the board, and financial reporters—and these groups' access to the public.

There was a gap between what the buyers saw as the reasonableness of their position and the best possible deal, and when the gap became known, instead of influential parties talking quietly behind the scenes, the buy-out of RJR/Nabisco became a plaything for the ever-inquisitive public, and image issues became central. Appearances—how the deal looked—were of the essence. The point is that the Johnson position appeared abysmal and unconscionable, because competing potential acquirers offered much more for the company than Johnson had pledged it was worth. These competitors struck pay dirt when *Time* published a cover story on the supposed naked greed of Johnson and his colleagues. When it became obvious that shareholders were being intimidated by Johnson, not rewarded, the would-be inside buyers lost the whole war.

Analyzing it now, we can see that while the ability to finance the transaction, the perseverance of the competing players, and the legal issues surrounding the RJR/Nabisco deal were of the essence, they do not entirely explain the outcome. To do that, one must have reference to the image issues. Shearson and Johnson simply miscalculated the image aspect, and, as a consequence, were publicly humiliated. Moreover, the principals' business careers were effectively destroyed, as Johnson was soon dismissed by the successful bidder, and Peter Cohen was dismissed as CEO of Shearson Lehman. So: in takeovers, how things look does matter, and anticipating and defusing possible scrutiny is mandatory. If, at the beginning of the saga, Shearson Lehman and Ross Johnson had correctly gauged the image factor in the LBO, they would never have offered $75 per share, never have structured a deal that would permit a handful of executives to split a billion dollars, never have agreed to excessive fees and commissions. These were all backroom agreements that would never remain unscathed in the merciless light of public scrutiny. Maybe if they had offered a better deal—a fair deal, one that reflected the reality of the situation—they would have won the day. But they didn't.

CHAPTER 17

For Tomorrow's Management

For today's most advanced corporations, the question is not whether to cultivate an image, since every company has one whether it wants it or not; it comes with the territory. Rather, the issue is what priority the leaders should assign to the image asset, and how it should best be managed in order to shape the most useful image to suit the company's long- and short-term goals. To understand image management, and how such a concept came into being, it's first necessary to summarize the sequence of steps the consumer goods industry in the United States employed. It was the pioneer industry insofar as image management techniques are concerned, and one which served as a model for all other industries, whether in the service sector or the manufacturing sector.

Stage One: Logo Solo

There was a time when establishing a brand image was a relatively simple task. In the nineteenth century, and for much of the first part of the twentieth, brand identities and logos evolved without a great deal of thought being put into them; many were people's names, or city locations, or obvious symbols of the products or services being offered. Ford automobiles. Wells Fargo express services. Bank of Omaha. Smith Brothers Cough Drops. Kellogg's Corn Flakes. Quaker Oats. Design experts had little to do with the process. Most companies would reach out to the "artistic community" to hire someone to make the brand look good; design, at that time, was an exercise in decoration, completely divorced from any business consideration. Up until the late 1940s, for instance, the Betty Crocker identity had been a General Mills fixture, but only as a name for a recipe giver. Then the company decided to do something more with its trademarked character. Lippincott & Margulies recommended that instead of limiting Betty Crocker's role to the "person" who responded to requests for recipes, she be given more proactive tasks. We suggested that she become a personality who would appear on packages, sales promotion items, and even in advertisements. We created a motherly, warm, All-American Mom look for her—an authority figure for the kitchen, and an implicit recognition that the way to a man's heart was through his stomach, a popular theory in the 1940s. (Some say it's still relevant in the nineties.) We also recommended that Betty Crocker be treated as a brand expression, through the design of the spoon logo in which she appears, and suggested that this brand image be separate from the corporate identity. For General Mills's corporate logo we created the now-legendary Big G, which could be used as a brand mark for products that were not marketed under the Betty Crocker identity.

Stage Two: Total Package Design

In the old days, prior to the widespread distribution of supermarkets, product information was conveyed to the consumer by the man or woman behind the counter of the local store. A package was then considered merely a container for the safe delivery of the product from the factory to the ultimate purchaser and user. In the modern era, as consumers began to pick their own choices off supermarket shelves—without benefit of advice (or salesmanship) from the proprietors of small local stores—this began to change. The first packaged brand was for Uneeda Biscuits, a generic name turned into a once-in-a-lifetime brand name. Its logo was a cross with two bars and an oval, supposedly an ancient symbol for the victory of good over evil, and the package itself had printed on it the sort of sales talk a customer might hear from a friendly grocer. The level of design was adequate because there was no competition, and the makers seemed to have an inherent understanding of what it would take for the package and the brand to succeed.

Simultaneous with the change from proprietor salespeople to packaging was the need for brands and logos such as

Betty Crocker to be used for more products than just cake mixes. As the problems of extending brands to larger product lines and finding ways to reach the consumer became more complex, there arose a need for greater involvement of designers in fashioning packages. Simple decoration was no longer an adequate response to the problems of marketing. The labels for Betty Crocker mixes, Campbell's soups, and General Mills cereal packages began to include important information for consumers, and also to act as advertising to convince the consumer to buy the product within that particular package, and not a competing product. Design became an as-yet-undefined, but clearly acknowledged, business activity. The recognizability of logos, the attractiveness of the package, and the messages that both conveyed took on new importance. Their continued visibility in the home became a significant factor in repeat sales, and in what we would now style brand awareness. While the function of the logos continued to be described as simply good design, there developed among the people involved in selling packaged goods the belief that all packages would now be required to communicate and persuade customers. "Packaging design" became an industrial specialty. (A side notion: If at the time packaging design began to grow, it had been renamed "packaging advertising," I believe it would have been awarded much higher stature and would have been able to command much larger fees—just another instance of how identity shapes image.)

Stage Three: Naming Brands

Beyond graphics and packaging, there evolved a need for more work on the words that conveyed the messages and shaped the images of brands. As more and more products crowded the supermarket shelves—and the new-car lots, and the advertising circulars of the Sunday papers—there had to be verbal differentiators in addition to logos and intriguing packages. Previously, names had been used for identification more than for any other purpose. Now, as new products were developed and the cost of properly introducing them began to soar, it became important to name a brand with an eye toward how that name would help sell the product; the product's identity would have to indicate more than just the manufacturer's moniker. Procter & Gamble diapers would not do—but Pampers was a name that signified the use and hoped-for result of the product; similarly, Duracell announced its qualities as a battery, Taster's Choice its aim as the coffee of the discriminating consumer. Henceforth, in addition to its other tasks, the brand name would help provide consumers with a reason for buying the product.

Stage Four: Identity Systems

From inventing logos, packaging, and names of brands it was a short but important leap to shaping the image of products and services over a broad range of media through the development of a total identity system. In such a system, all of a corporation's communications and its flagship brands were rigorously scrutinized and regularized to reinforce a single identity. Stationery, signage, packaging, logos, and every other visual expression were understood to be related to the corporate or brand identity, and, therefore, were in need of being aligned with the approved objectives. The prime example, here, was Lippincott & Margulies' organizing of the communications of Coca-Cola. Later on, the same need arose for companies that provided more services than products, as exemplified by our work for American Express.

Designing a logo that conveys the proper symbolic message, creating packages that help sell and inform, developing names that reflect the intent of a product or a service, even creating entire identity systems, are things that today seem obvious and necessary for any large business to do. Yet at the time these innovations were first introduced, they were not instantly understood, and their high level of acceptance today is evidence of their importance and good fit with American business. Perhaps more to the point, once most companies had begun to devote considerable effort to these identity-related notions, it became imperative for any company seeking to introduce a new brand in the marketplace to go through these four stages to establish its identity. Today, as opposed to years ago, many of the stages are combined and simultaneous, so that a new product comes into the world replete with an organized and effective name, logo, package, and associated identity system components firmly in place. If it didn't, there might be no sense in even bringing it to the public.

And so we have arrived at . . .

Stage Five: Image Management

Although the need to maintain and to continuously manage an image should be obvious to all corporations, I am sad to say it is not. Some do it very well; others manage their images only at the introduction of products or services, and fail to give them attention at any other time. Still others recognize the need for image management, but abdicate their own responsibilities in this area and let advertising agencies do it for them, even though the use of advertising is only one of the many aspects of managing images. A fine car can continue to run for many years (and give good service to its owner) if it is regularly examined, tuned, and cared for, but if the oil is not changed at proper intervals, the tires not rotated, nor the engine investigated for hidden problems that can then be addressed, even the most expensive and beautiful vehicle can come grinding to a halt and end up useless. Corporate images need similar constant attention if they are to render good service (and continued value) to their owners.

A recent incident involving Continental Airlines underscores the crucial importance of the issue. Last December, when Continental Airlines petitioned the court for protection from creditors under Chapter 11, the airline included in its appeal a plea that would allow it to continue funding its image upgrade program. According to the affidavit, Continental maintained that "the centerpiece of our five-year plan for survival is an accelerated wholesale overhaul begun months ago to upgrade our image." With this statement, management proclaimed their belief that Continental's existing image was impeding the company's survival. Releasing the funds, their argument reasoned, would benefit the airline's creditors and shareholders because it would enable the company to implement programs that would substantially change Continental's image and increase revenue. Although a decision on this plea had not been handed down when this book was written, management anticipated that the court would approve this aspect of the petition.

Continental's circumstances aside, too many business leaders deem image management to be a highly discretionary activity, an expenditure all too easy to postpone. In fact, quite the opposite is true. Continental's seeming indifference to its image caused the airline many problems—many crippling ones at that. Obviously, Continental's financial difficulties cannot be blamed only on image problems; however, they unquestionably contributed mightily to the airline's sad state of affairs. To Continental's credit, it began the process of correcting the problem a year before several unrelated events forced the company into Chapter 11. Continental's appeal to continue its image program—and (hopefully) the court's allowance—exhibits a growing recognition of image management's central role in business. The old school of management viewed the subject of image as no more meaningful than the cost of a fresh coat of paint on an airplane. Today's enlightened executives are according image management the recognition its merits as a vitally important and basic business tool.

* * * * *

In the previous chapters, I hope to have demonstrated conclusively that many important and complex business problems can be addressed, understood, and sometimes solved through the special prism of communications. Now, in summary, I'd like to present some recommendations and admonitions for tomorrow's management, in light of what I also hope to have established as communications imperatives.

The Essence of Corporate Leadership

In the most recent annual survey conducted by *Fortune,* a poll of eight thousand high executives, outside directors, and financial analysts ranked the most- and least-admired public companies. The most-admired companies in 1989 were Merck, Philip Morris, Rubbermaid, Procter & Gamble, 3M, Pepsico, Wal-Mart, Coca-Cola, Anheuser-Busch, and Du Pont. It surely is no accident, then, that the most-admired executives included such names as Lee Iacocca of Chrysler, Sam Walton of Wal-Mart, Robert Goizueta of Coca-Cola, and Dr. Roy Vageles of Merck. But what I want to point out is that these are precisely the corporate executives who are the most clearly cognizant of the relationship of image to corporate health.

The survey also demonstrates another tenet of mine, that the primary component of a good image is good performance. While the most-admired companies were rated on many different criteria, one statistic stands out: the average of their total return to stockholders for the past ten years was 26.58 percent annually. In 1989, the rise in the price of their stock was a whopping 47.4 percent. And just compare the financial performance of the ten least-admired companies, namely, Gibraltar Financial, Wang Laboratories, Control Data, Meritor Financial Group, Texas Air, CTV, National Steel, United Merchants and Manufacturers, K-H, and Unisys: their average total return for the last ten years (using only seven of the ten because of incomplete records) was 10.48 percent, and the decline in their stock prices for 1989 alone was a dreadful 37.9 percent.

While these statistics relate to the very top and very bottom of the curve, the general lesson is crystal clear: corporate image is a basic reflection of the state of corporate health, and image and corporate health are intertwined. And so we deduce the corollary: *The good corporate leader is one concerned with the corporation's basic health and who simultaneously assigns a high priority to the significance of its image.* The successful corporate leader recognizes, too, that the subject of image is never something to be considered in a vacuum, or separately from other aspects of the corporation; nor can it be considered a rabbit to be pulled out of a hat to distract the public when troubles loom. Rather, the maintenance of a corporate image is the result of a disciplined, long-term attitude and policy.

Interestingly enough, those CEOs who deal badly with image management are invariably those whose leadership is reputed to be dictatorial rather than consensual, and who do not believe in nor try hard to build the feeling of teamwork among their management troops. Such CEOs are adamant that performance can be and should be measured solely in terms of absolute profit, and they demonstrate a single-minded, conventional focus on "making their numbers." Many of these chieftains are not true leaders because, though they may command obeisance, they fail to inspire their employees. Insensitive to other people, they refuse to take into account anything they have not personally understood; when they are ignorant of something in business, they ignore it!

Often, one finds them publicly railing against the short-term mentality of Wall Street, which, they say, forces them to do almost anything to increase earnings every quarter; they cry that they really are oriented toward long-term growth—but then they don't pay much attention to the firm's image, which is a definite long-term matter. Or they profess concern with image, but won't spend very much money to bolster that image, or to alter it if change seems imperative.

Having met hundreds of CEOs in the course of working in the field of corporate identity and image management, I conclude that real leaders—those who have the true support of their troops, and can command by the power of their vision—are *always* conscious of image; moreover, they closely link image with reality, and never consider them as separate entities.

Leaders are those among us who develop and articulate visions of the future. Some have vision, but nothing beyond it. Others—the more capable ones—also supervise the way in which this vision is translated into reality, providing the day-to-day oversight and encouragement to their followers engaged in the process of creating the desired future. It is another attribute of leadership to orchestrate the musicians and motivate them to keep playing. Now the key point I want to make about these three aspects of leadership—*vision, oversight, and motivation*—is that they all have to do with communications. Since the end result of communications is the shaping of an image, a central imperative of leadership can be said to be image management.

Not everyone will go as far as I do in this contention, of course, but even critics must admit that today, at the very least, leadership and image management are interconnected concepts. Those CEOs who have understood this intuitively, or through the intercession of outside help, have succeeded, and those who have not—like Rawls in the recent Exxon crisis, or Johnson in the RJR takeover fiasco—have failed to weather the crisis without harming the company, or have even lost control of their empires.

So I say that a primary task of a CEO is image management—today, more so than in the past, and tomorrow, increasingly so. But it is not a task that can be reserved only for CEOs. Protection and enhancement of the company's image is one of the few aspects of business that embraces all of a company's employees. What the employees do, how they perceive the company and represent it both to the outside world and to one another, all have a direct effect on image. The CEO who will demonstrate his belief that burnishing the company's image is a priority will produce followers who will also work hard toward this goal. Consider the success of Sam Walton. The founder of Wal-Mart is one of the country's richest men; he has established a requirement that all customers must be personally greeted when they enter a Wal-Mart store, because he believes that such personal service will bring customers back time and again more surely than will cut-rate prices. To my mind, Walton's requirement—his concern with image management—is an integral element in his company's enormous vitality.

If you were to ask Bill Gates of Microsoft if he spends much time burnishing his company's image, he would probably look at you as if you were nuts—but that's exactly what he is doing by responding to all memos and phone calls from his executive staff on a daily basis, by tirelessly communicating, by insisting on clarity and precision in communications, by being devoted to his company and its people. In nourishing the reality by strengthening the staff, you reap a double benefit: a better staff, and a better image.

On the other hand, the CEO who is not concerned with image will likely lose an important link between the employees' efforts and the company's fortunes, with potentially dire consequences. When the Pillsbury Corporation correctly feared it was about to be the target of a takeover bid, it asked an outside consultant to evaluate its strengths and weaknesses. One problem uncovered in this study was that Pillsbury's valuable Burger King franchisees did not feel well tied in to the corporate management; similarly, a survey of the company's employees revealed their sad belief that the corporation did not care much about them. Despite the consultant's warnings, these problems—which I would definitely label aspects of a faulty corporate image—were ignored. Shortly, when the takeover bid materialized, Pillsbury executives were unable to marshal internal support from their Burger King franchisees or from their larger group of employees, and the company fell to the raider, GrandMet, within two weeks.

Good leadership understands that image management and concern for image can and should be fostered. When celebrated, the company's image can become the medium for the corporate culture, the flag to which everyone rallies. Human resources experts agree that when a company has a good and strong corporate culture, it is able to attract employees easily, to generate loyalty to the company among all employees, and to inspire them to do the best job possible. In a company that manages its image well, employees will believe that the company is a good place to work, cares for its workers, and recognizes their best efforts. Producing such feelings in the hearts of employees is also a consequence of good leadership and of good communications. That is why good leaders watch their internal communications as carefully as what they say to the outside world. Do all the corporation's communications reflect a certain style that is easily identifiable and of appropriate quality? Is what is being said easily understandable and credible? Do internal communications seem organized? Is there a feeling that top management cares how the rank and file feel about them? Is it clear that management recognizes that there is more to a job than a paycheck, and that these intangibles are significant? The answers to these questions can be found in internal communication practices that are a vital part of image management.

The leader must not only wave the flag, he must lug it up the hill. If he does so properly, the troops will charge right alongside him, and carry the day.

Tasks of an Image Manager

In the far corporate past, a relatively low-level manager had charge of the company's personnel. Then, as the corporate world grew more sophisticated, and as the problems of hiring, firing, and maintaining employees grew more important and more complicated, the lowly personnel manager emerged as the Director of Human Resources, a position in which he or she reported directly to the highest levels of management. Today, image management is more important than ever before perceived, and the time has come to recognize this within corporate hierarchies. What will this mean to tomorrow's companies?

To coordinate the images that a corporation is presenting, in the future I believe companies ought to have a central department, or at least someone at headquarters who has central responsibility, for image management. The job of this department (or of the vice-president) would be to establish image objectives on a corporate and a marketing level, and to then manage the company's images in ways that are both consistent with the objectives and cost-effective. Supervised by this department would be all of the corporation's corporate identity: its communications, internal and external; the direction of its advertising and public relations; its investor relations; its sales promotions; and perhaps even the company's merchandising efforts. All of these need coordination, and this department would handle that task. Tone and manner of communications would be preserved, so that two departments such as business long-distance and residential services would never unwittingly mount contradictory advertising campaigns in the same medium.

Let's suppose a company had a vice-president who had total responsibility for image management for all of the company's endeavors. What would be the tasks of an image manager?

Relevance

The image manager must work to make the company's image, and that of its branded products and services, relevant to current conditions, and, where possible, strive to make it timeless. In terms of an overall company image, this is a difficult task because there are so many elements that go into the mix. Chrysler's fortunes were at a low ebb until Lee Iacocca decided to take the company in hand. The common perception is that Chrysler was floundering helplessly until Iacocca arrived. Its problems of quality control, performance and styling, the relationship of the company to its dealers, and its balance sheet were well known—broad awareness of these problems was in itself a huge crisis for the company. As the new leader, Iacocca mounted a coordinated attack on the company's structure and finances, its relationships with suppliers and customers, and its reputation with its various audiences. By trumpeting to the public the challenges facing Chrysler and the ways in which the company was fighting to stay alive, an old, sick company was made to seem heroic. People began to care about what happened to Chrysler, and this increased its relevance. Thus, while the reality undoubtedly had to be corrected for a real turnaround, without careful attention to image issues the battle simply could not have been won.

In the instance of Chrysler, the company was rejuvenated, and so was its image. Similarly, a company's product can be altered, or the messages conveyed about it can be changed.

Consider, in this regard, the four best-known consumer products to come out of World War II, those that all GIs seemed to use: Jeeps, Coca-Cola, Camel cigarettes, and Zippo lighters. In the ensuing years, Jeep managed to change the engineering and styling of its products to assure them of new buyers. Coca-Cola did not change the product (except once, for a few months), but continually readjusted its advertising to maintain Coke as a contemporary drink for that most fickle of audiences, teenagers. Camels tried with filters but could not alter the basic product or its image; once, RJR attempted to alter Camel packaging, changing the number of pyramids on the pack, and suffered a decline in sales, after which the change was withdrawn and further innovations discouraged. This experience alone was enough to intimidate anyone in the organization who sought to artistically modify Camels. Today, of course, cigarettes are becoming less salable in the United States, so Camels are in decline, but still viable, thanks to their past history and to changing advertising campaigns. In the late 1940s Zippo lighters were ubiquitous, and Americans used them (and almost nothing else) to light their smokes; Zippo was dominant in the marketplace. Then competition rolled in from Bic and other cut-price lighters, but the Zippo company chose not to meet the challenge in any aggressive way, steadfastly refusing to change either its product or the way it was marketed. Zippo remained frozen in time, both its products and its image. By refusing to change anything in a world that was itself rapidly changing, Zippo suffered the consequence: a shrinking business.

Banks used to be thought of in terms of their trustworthiness and the convenience of their locations. Those images served many banks quite well until technology and deregulation outmoded and irretrievably altered the perception of what was necessary to project the trustworthiness factor, or the value to the customers of convenience. In the era of ATMs, who needs a convenient branch bank? The terminals are everywhere, and networks of them, such as Cirrus or NYCE, allow depositors of one bank to withdraw funds from the physical plant of another, and even from the ATM machine in the corner supermarket. Trustworthiness, while always important, is almost taken for granted because of government insurance of deposits. Certainly, it is not necessary for banks to be housed in marble palaces featuring tellers behind wrought iron bars to communicate the safety of a deposit in the bank. Curiously, while the banks have adapted to the new era, many have not made their facilities or their images contemporary enough: they still design new banking centers as if serving people through teller windows was a bank's main function, and in their communications they still emphasize the same old story, instead of, for instance, the convenience and availability of electronic money transfers.

Sharp Focus

The vice-president/image manager must take pains to insure that the company's images do not become diluted or lose focus. Loss of focus can come as a consequence of trying to make a single brand encompass too many products. When Hershey Foods bought San Giorgio Pasta Company, should the Hershey image have been extended to the pasta products? Should Campbell's soups have married its identity with that of its subsidiary, Lady Godiva Chocolates? In both cases, the answer was clearly no. A prime example of an image being extended too far is Cadillac. Cadillac was once solely synonymous with luxury cars. Then, in the past several decades, the luxury label was assiduously courted by BMW, Mercedes-Benz, Jaguar, and other foreign carmakers who expanded their sales into the United States, and Cadillac lost its preeminence in the luxury field. In an attempt to gain more sales, the makers of Cadillacs decided to offer the Cimarron model, a less expensive car that carried the still-powerful Cadillac label—a maneuver that proved to be a mistake, because the existence of a less expensive model undermined further the luxury image of the Cadillac line, and would-be buyers chose competitors' cars in greater numbers than ever before. A good image manager would have understood that sharp focus demanded that a luxury image not be diluted by extending it to nonluxury items. To state it another way: pricing policy not only affects profits and losses, but also, a product's image.

Consider what happened to Xerox, the corporation that

pioneered with brilliance, innovativeness, and single-minded focus the office copying field. In the 1960s when we recommended to the then-named Haloid Xerox Corporation that it shorten its name to the Xerox Corporation, the chairman vigorously resisted the idea; he felt that Xerox was such a strange and alien-looking-and-sounding name—difficult to pronounce —that it would discourage customers from buying the product. Ironically, one of the problems that the company faces today is that Xerox has become so easy to pronounce, and so close to being a generic word—a verb, in common usage; to become a generic is generally a death sentence for a brand image, unless the image is carefully managed. To complicate the picture, in the 1980s the company began to develop new business lines in total office automation and financial services, and identified these as Xerox products. It hoped that Xerox could be redefined to stand for excellence, rather than for a specific product. But it became a victim of its own success. In the mind of the public, Xerox can stand only for copying machines, and the attempt at redefinition dilutes the power of the name; as a consequence, the company's image is unfocused.

Still another example: Merrill Lynch's entry into the residential real estate brokerage industry. Merrill Lynch had successfully expanded its excellent image beyond the selling of stocks and bonds that formed the foundation of the enterprise, to embrace a range of financial services and products; the thread that connected all of its activities was the company's ability to provide well-trained professionals to manage their clients' financial assets in an organized and effective manner. It must have been quite tempting to management to try and get involved in what for many of its clients was their largest single asset—their homes. Since the business class that constituted the bulk of Merrill Lynch's customers was increasingly mobile, real estate brokerage sounded like a surefire idea. Initially Merrill Lynch acquired small mom-and-pop real estate brokerage firms around the country and linked them nationally through computer systems, affording these once-small regional firms a competitive edge against local firms and allowing Merrill Lynch to go head-to-head with established national residential real estate networks such as Coldwell Banker and Century 21. An added asset for the real estate division was the well-known and trusted Merrill Lynch name and image.

However, the venture was just not for them. While it is commonplace to say that success has many fathers and failure is an orphan, this particular failure had distinct parentage. An image must stem from a company's reality, and one must match and reflect the other. The foundation of Merrill Lynch's image was the professionalism of its thousands and thousands of corporate brokerage employees, something emphasized in the corporate culture. Unfortunately, in the residential real estate business, most employees are not full-time. They were part-time free-lancers, randomly trained, and uncommitted to a particular corporate culture. While the Merrill Lynch name and image could open doors for these real estate brokers, once the door was open, customers soon realized that there was no further inherent advantage in having a Merrill Lynch real estate broker rather than a Coldwell Banker broker or an obscure local one—and the real estate brokerage business did not become as profitable as Merrill Lynch would have liked. Eventually, top management realized not only that the venture itself was of questionable financial value, but also, and just as important, that it could inflict a lot of damage on its most valuable asset, the Merrill Lynch image.

Things quickly went downhill. The enterprise was offered for sale, with the stipulation that the Merrill Lynch name could no longer be used on the real estate brokerages. The response by potential buyers? No name, no sale. Eventually, the entity was taken public, under the name Fine Homes, but marketing continued under the Merrill Lynch identity because, in fact, Merrill Lynch still owned the majority of the shares. Eventually, the entire operation was sold to Equitable Real Estate, and the Merrill Lynch name was retrieved and removed from the real estate brokerage field. While lasting damage was avoided—at quite a cost—it is possible that proper image management would have anticipated this series of events, and avoided the problem in the first place.

Positioning for the Future

Image management's responsibilities include positioning the corporation and its products for the future. For example, Macintosh computers are billed as the only computers a purchaser will ever need, because Apple is pledged publicly to use current systems as the basis for future improvements in the form of add-ons, rather than total replacements. Its identity systems facilitate this strategy. More usual is the need for a company and its products to compete globally rather than only in an American marketplace. Is the corporation's image too provincial? In order for a bank in Seattle to increase its base throughout the northwest region, it became Rainier Bank, named after the geographical area's most notable feature, but avoiding the limitations normally associated with geographic names. In the future, most experts agree, large shares of the international markets will be wrested out of the hands of regional marketers by global brands such as Sony, whose name and image, though Japanese, easily traverse national boundaries. The era of "passportless" brands is upon us.

Will the product names travel well? Ford began manufacturing its Pinto model at its Brazilian plant and intended to market it in South America; however, when assembly line workers roared with laughter as they saw the nameplates being attached, management suspected that something might be wrong. It was. In Brazil, the slang definition of Pinto was small male genitals; the word was also used in Brazil to describe a husband's inability to satisfy his wife sexually. The Pinto in Brazil became Mustang II.

The value of a brand can go up or down, and should not be taken for granted, deemed to be an invariable constant or thought to be infinitely elastic. Take the case of Seagram's, a premier purveyor of hard liquor for many years. Today, revenues from hard liquor are declining as consumers switch to wine, wine coolers, and softer drinks. The corporation has announced that it is aiming to become the world's best beverage company, and toward that end has acquired Tropicana Orange Juice and Soho Natural Sodas, and brought to the fore the sales of its own Seagram brand of ginger ale, club soda, and other soft drinks. Is the Seagram's identity elastic enough to encompass these softer beverages? Or should the company build other identities for them, in order to rid such products of the perhaps-unwanted association with hard liquors? This is an area in which the image manager would have a lot to say about a company's future directions.

Anheuser-Busch had a problem similar to that of Seagram's, and solved it through remarkable astuteness and foresight. It developed equity in several of its businesses by building up brand identities that were separate but yet connected to one another in subtle ways. The Budweiser brand is only one of these. There are also Eagle Snacks—a concept that evolved from the eagle in the corporate logo—and Busch Gardens, to name just two thriving entities. They seemed to have found a way to have their beer and drink it, too.

Resource Allocation

Surely one of the most important of the VP/IM's responsibilities is to determine whether the corporation's resources are being properly spent on image matters, and to oversee those expenditures. Reporting to that VP would be people in charge of advertising, public relations, internal communications, shareholder relations, and so on. Acting in concert with these subordinates, it would be imperative for the VP/IM to take decisions about allocating company resources out of the hands of outside vendors. Currently, many such decisions—spending levels, types of campaigns, which aspects of an image to bolster or let slide—are made by advertising agencies. In a typical scenario, the VP for advertising asks his or her current agency, or several competing agencies, for suggestions for the coming year's expenses and campaigns. But the agencies' incomes are linked to a percentage of every advertising dollar spent! This process of asking agencies for estimated budgets effectively usurps the power of company officials to make decisions and encourages the agencies to make them—when the real questions to be pondered are more fundamental and more important: should there be any advertising next year, and, if so, what kind and where. The ad agencies—no dummies—have recognized this usurpation and have attempted to defuse the issue by acquiring subsidiaries that have various specialties, and suggesting to clients that all of the communications options be considered by the umbrella group. To me, this is like the aggressor asking the victim if he would like to be hit by the aggressor's right hand, his left hand, or one of his feet. Since the bottom line of an umbrella agency's balance sheet is driven by the advertising budgets of its clients, I do not believe that agencies can be truly objective when it comes to deciding whether or not their clients should advertise. Old habits die hard. Furthermore, the theory put forth by many agencies with specialty subsidiaries—that one-stop shopping is a boon to the client—collapses under careful analysis. Can a client's image management needs be best served by a department-store type agency that provides all possible services to the client, regardless of the excellence (or non-excellence) of its component departments? It seems to me that clear-minded, objective strategic thinking and creative excellence is much more important to a client than one-stop shopping. While I understand the advantage to the purveyor of such services, I don't see as many advantages for the buyer. Related service companies have no greater motivation to succeed with a client than do independent ones.

Add to the problem of budgets and allocations the task

of deciding what media to use. In the future, we are going to be communicating messages in many more media than we are using today, and the coordination of those messages and the media in which they appear presents complex problems. Electronic vs. print? Global vs. regional? Sustaining vs. short-term? These are proper questions to be asking, and all the more reason for having a VP/IM to investigate and decide them, rather than simply turning them over to an advertising agency.

A New Director?

As image matters take on an increasingly weighted priority in the day-to-day business picture, they will force companies to assign high-level employees to deal with image management. In suggesting a VP/IM, I hope that I'm advocating the inevitable. But the change wrought by an understanding that image is a valuable company asset must be recognized (and championed) even further up the line. Who on the board of directors is going to have oversight responsibilities for image matters in the future? Most corporate boards tend to be loaded with bankers, lawyers, academics, former government officials, and company presidents. Very few of these people are qualified to evaluate the key decisions that go into the managing of images. Perhaps someone expert in the area could take that next vacant director's seat? Do you really need another banker on your board, or would it be better to have someone who can make a more unique contribution to the corporation's taxable income?

Let's go a step further, and talk about top officers of the corporation of the future. If, as I have argued, the corporation's image is one of the company's primary assets, why should the only people considered for the inner executive circle be those whose experience has been in line management, legal dexterity, or the like? Going beyond appointments entirely, the image-management sensitive company of the future will do something else not currently done: institutionalize the importance of image management. This will happen when tomorrow's companies recognize, celebrate, and promote those people within the corporation hierarchy whose perspective and particular skills enable them to manage the company's image in a manner appropriate to its significance and impact on profits, losses, and future strategies.

The Image Decade

For businesses, especially those large, public companies committed to steady growth, the 1990s will be the image decade. Companies wishing to hold on to their positions of leadership, or those who want to seize the leadership, will come to realize that image (as an accurate reflection of reality) is among the most important assets of the company, and in some instances the company's major asset, principally because it is wholly proprietary. In the image decade, globalization, the multiplicity of communications, and ever-larger budgets required for marketing will be the watchwords for all companies. Those that seize the moment best will be the organizations that accord to image-making and image-maintenance activities the status, money, and proportion of top executives' time that such issues deserve. These leaders will be the companies who understand that as we cross into the twenty-first century, image-related activities will only become more and more important, and will require more sophistication, planning, and effort. They will be the companies who first understand the complexity of the new technologies in which messages can be communicated, and exploit the advantages that their advanced understanding affords. Taking into account the financial consequences of their image-related decisions, they will recognize that, once and for all, a company's images and identity practices will have to be managed differently from the way they have been in the past, for there is no going back to the old ways—but they will have little fear of what tomorrow holds, secure in their understanding that by controlling their images through conscious design, they will be in the best position to take advantage of the future.

Image never exists in a vacuum but is integrally related to reality. Image is a gauge for assessing the company's health, as telling a reflection of the company as return on equity, profit margin, or earnings per share. Image must be designed because business, in its everyday concern to improve its own performance, must use every tool at its command to assess its strengths and weaknesses, to compete, and to prevail in the hard and competitive world of commercial reality.

Marsh & McLennan Companies, 178
MasterCard, 70
Mead Corporation, 63
Medium as message (McLuhan's concept), 149–150
Mercedes-Benz, 146
Merck, 196
Mergers, 42–43
Meritor Financial Group, 132, 196
Merrill Lynch, 126, 177, 203–204
Metropolitan Life, 93
Meyer, Edward J., 14, 16
MGM, 154
Mickey Mouse, 92–93
Microsoft, 198
Milken, Michael, 176
Miller beer, 29–30
Miller, John, 41
Minkow, Barry, 57–58
Minority workers, treatment of, 17–18
Mitsubishi, 7
    control of Rockefeller Center, 7
Morale. *See* Employee morale
Muppets, 93

Nabisco. *See* RJR/Nabisco
Naming, 52–53, 129–141, 193–194
    of brands, 193–194
    of business or service, 129–141
    dictionary vs. fabricated names, 132–134
    effect of name over time, 136
    line extension, 139
    names as unifiers, 134–135
    nomenclature architecture, 137–139
    practical considerations, 130–132
    image management and, 20–21, 25–26.
    See also Logos
Napalm, 50
Narrowcasting, 149
National Can Co., 72
National Steel, 196
Navistar International Transportation Corporation, 14
NBC, 70
New England Telephone, 98
New York Life, 39–40
New York Telephone, 96, 98
Newark, New Jersey, 5
Newman, Paul, 91
Newsletters, 176

Nissan Co., 146
    introduction of Infiniti auto, 99–103
No Excuses, 88
Nomenclature architecture, 137–139
Norden military devices, 167
NYCE, 202
NYNEX Corp., 3, 14, 98–99

Ogilvy & Mather agency, 80, 163
Otis elevators, 167

Pacific Car and Foundry, 133–134
Package design, image management and, 192–193
Paine Webber, 43
Pampers, 130, 137
Paramount Communications, 169
Parenting. *See* Corporate parenting
PBS, 70
Pentastar symbol (Chrysler), 74–78
People's Express, 131
Pepsico, 48, 196
Perception,
    of business or corporation, 22–36.
    See also Image management
Perelman, Ronald, 42, 185
Perfume industry, Image management in, 87
Perrier water, 27–28
    tainting of, 27–28
Personal computers, 151
Pet Milk Co., 133
Peters, Jon, 184
PetroCanada, 126, 133
Pfeiffer, Michelle, 90
Philip Morris, 162, 179–180, 196
Philips Gloeilampenfabrieken, 131
Pickens, T. Boone, 183–184, 186
Pillsbury Corp., 163, 199
Politics, image making in, 5
Post-disaster syndrome, 44–48.
    See also Crisis situations
Pratt & Whitney, 167
Premiums, 67–69, 71–72
    aspects of image in, 68–69
    defined, 68
    the "no premium blues", 72–73
    See also Commodities
Press releases, 176
Pricing. *See* Costs
Primerica Corp., 67, 109–111, 132, 174